Birds Asleep

NUMBER FOURTEEN
THE CORRIE HERRING HOOKS SERIES

BIRDS ASLEEP

Alexander F. Skutch

ILLUSTRATED BY N. JOHN SCHMITT

UNIVERSITY OF TEXAS PRESS AUSTIN

Printed in the United States of America

First edition, 1989

Requests for permission to reproduce material from this work should be sent to:
Permissions
University of Texas Press
Box 7819
Austin, Texas 78713-7819

LIBRARY OF CONGRESS CATALOGING-IN-PUBLICATION DATA

Skutch, Alexander Frank, 1904–
Birds asleep.

(The Corrie Herring Hooks series ; no. 14)
Bibliography: p.
Includes index.
1. Birds—Behavior. 2. Birds—Physiology.
3. Sleep. I. Title. II. Series: Corrie
Herring Hooks series ; 14.
QL698.3.S54 1989 598.2'51 88-33954
ISBN 0-292-70773-8

∞ The paper used in this publication meets the minimum requirements of American National Standard for Information Sciences—Permanence of Paper for Printed Library Materials, ANSI Z39.48-1984.

To Paul Feyling

Contents

Preface

Diurnal birds spend about half their lives sleeping, or at least resting quietly in their roosts or dormitories. This is most obviously true in equatorial regions, where day and night are approximately equal in length throughout the year. At higher latitudes in summer, where days are longer than nights, birds are active for the greater part of the twenty-four hours; but long midsummer days are balanced by long midwinter nights, with the result that birds who reside permanently at these latitudes pass about half their lives resting. By spending the northern winter in the tropics, birds that nest in the temperate zones may increase their yearlong total of waking hours. Birds who, after nesting in the Arctic where daylight is continuous in summer, migrate to high southern latitudes enjoy the greatest amount of daylight in the course of a year; but they take some rest, even in the land of the midnight Sun. Indeed, all animals appear to need sleep after activity.

The ways that birds sleep are exceedingly diverse. Many roost in trees or shrubs or marsh vegetation; some sleep on the ground, others floating upon water; some lodge in holes that they carve in trees, burrows that they dig in the ground, or, especially in the tropics, covered nests that they build for sleeping. A few birds sleep, or take whatever rest they need, high in the air. Birds may sleep singly, in pairs, in family groups, or in communal roosts with many thousands of occupants. The most careful parents install their fledglings in a sheltering dormitory. The sleeping habits of birds are so intimately related to their social life that knowledge of one may shed light upon the other. To find a safe place to sleep is no less important to birds than to find a safe site for their nests. In regions where winter is severe, as also on tropical mountains above the timberline where the night air may be frigid, birds must be careful to choose spots where they can maintain their body temperature without exhausting their reserves of energy. To them, as to us, freezing is fatal.

The great diversity of the sleeping habits of birds makes their study fascinating. Certainly, to understand how they live we must learn how they pass the darker half of their lives. Nevertheless, students of birds, both professional and amateur, have given less attention to this facet of bird life than to many others, such as food and foraging, reproduction, voice, molt, physiology, and migration. The general neglect of the ways birds sleep is easily explained. In the first place, with few exceptions, including those that roost in large and noisy crowds, it is not easy to discover where birds pass the night. Diurnal birds—the vast majority of avian species—are at a great disadvantage in the dark. Unlike mammals, which depend so largely upon scent, they guide themselves almost wholly by sight, which is of all senses the least effective in the absence of light. Accordingly, as they go to rest when night approaches, they are careful to avoid being seen by creatures that might harm them. They are at least as cautious in approaching their sleeping places as in approaching their nests. And as they go to rest in the evening, the failing light is their ally, making it more difficult for us to follow their sudden, swift movements with eyes more efficient in brighter illumination.

The best way to learn where and how birds sleep is to watch them as they retire at nightfall or become active at daybreak, but these events often occur at hours inconvenient to birdwatchers. At middle and high latitudes in summer, the student of sleeping habits must remain up and about late in the evening, when one might prefer to eat, read, or rest, or arise betimes in the morning, curtailing sleep. In winter the hours for making pertinent observations are much more favorable, but the weather is frequently adverse, chilling one who stands motionless for many minutes, patiently waiting for a bird to enter its dormitory in the evening or to emerge in a blustery dawn. In the tropics, except at high altitudes, these inconveniences are avoided: one need neither arise so early in the morning nor remain out so late in the evening as in summer nearer the pole, nor feel one's fingers grow numb while holding a binocular or trying to write notes, as in winter. The chief impediment to observation may be a tropical downpour. Moreover, because of the variety of nests that tropical birds build for sleeping, and the diverse ways they occupy them, the birdwatcher in the tropics is stimulated to strenuous exertion to uncover their secrets.

For more than half a century, I have spared no effort to learn how birds sleep, above all in tropical America. In addition to telling what I have discovered in the field, I include in this book such published information as I have found on how birds in other regions sleep. Too often those who write about birds give fascinating glimpses of sleeping habits without reporting all that I need to know to place the birds in the proper chapter in this book. In arranging my material by sleeping habits rather than by taxonomic categories such as families or orders, I have sometimes been perplexed as to just where an observation should go. Some birds would fit as well into one chapter as another; certain species appear in several chapters. The reader interested in a particular species might consult the index, which ties scattered references together. In it I give the scientific names of all organisms whose English names are capitalized in the text, since it is easier to find the scientific name of a certain species in the index than it would be to rummage through the text for the page where the species is first mentioned.

Studying how birds sleep may not be as exciting as watching them in full daylight when their colors are most vivid, listening to their songs, finding their nests, or making lists of them. But to know how they pass the more obscure half of their lives is necessary to round out our picture of their habits. This knowledge brings a sense of intimacy that is deeply satisfying. And this, as far as I know, is the only book devoted wholly to how birds sleep.

ONE

Oceanic Birds

The ways that birds sleep are no less diverse than their ways of foraging, nesting, or any other aspect of their lives. While some species work hard to prepare snug shelters for themselves, others settle down at nightfall wherever they happen to be, apparently regardless of comfort and safety. Let us begin with birds whose manner of sleeping appears the most primitive because they rest upon water or earth with little or no preparation for their repose.

One stormy November, many years ago, I crossed the turbulent North Atlantic from west to east on a slow freighter. On all the twelve-day voyage, seabirds were in sight of the ship throughout the day, and even a few small birds of the land—a Dark-eyed, or Slate-colored, Junco and a European Starling—borne out to sea by adverse winds, sought refuge on our vessel.

After we left the urchin House Sparrows that swarmed around the docks of New York, Herring Gulls began to follow us down the Hudson and through the bay. For three days they continued to accompany our ship, ever alert to snatch from the water anything edible thrown out from the galley or stirred up by the vessel's passage. But long before we reached the mid-Atlantic, they disappeared. Frequenters of inland and coastal waters rather than the high seas, our Herring Gulls probably did not follow the ship so far that they could not rest on a rocky islet, a sandy beach, a marsh, or perhaps on the water of some sheltered cove, pond, or river. Sometimes they roost in trees; and on the Isle of Lundy, Richard Perry found them resting through winter nights on the exact spots were they had laid their eggs in the preceding summer and would lay them again in the coming season. Great Black-backed Gulls, of which we saw the last as we passed beyond the Grand Banks, likewise rested on the sites of their nests.

After the larger gulls fell behind, Black-legged Kittiwakes appeared

Black-legged Kittiwakes on stormy sea

in increasing numbers around our ship. From the Grand Banks to Fastnet Light off the southwestern end of Ireland (where we again met Herring Gulls, now of the European race), we were each day followed by a company of these dainty, graceful little gulls, either circling in wide arcs around the vessel or soaring on spread wings in the ascending current of air on the windward side, to drop instantly to the water whenever the ship's cook threw out scraps of food. Through all the short, bleak days, the hungry kittiwakes stayed close to the laboring vessel. But as dusk began to settle, at this season and latitude about half-past four, they would glide down in numbers to the surface of the ocean, folding their pointed wings about their bodies and floating lightly on the waves, face to the wind, like anchored ships in a breeze.

The birds were at this hour torn between two conflicting impulses, to go to rest in the fading daylight, and to follow the ship in anticipation of retrieving yet another morsel to help tide them over the long, stormy winter night. Again and again, after settling upon the water as if to sleep, they rose once more to join others who still hovered around the vessel, waiting for something to eat. But as twilight deepened, more and more sank down upon the billows and grew rapidly smaller in the wake, until they were lost to sight upon the dark water. The latest to retire continued to accompany us until the light became too dim to distinguish their forms, except in silhouette against the darkening sky. But before daylight had wholly faded, the last had settled upon its cold and watery bed. There we left them, literally rocked in the cradle of the deep, brave little craft strung over miles of the open sea, riding out the gales of a stormy winter night on the North Atlantic.

Off the coast of Nova Scotia, in a sea very rough after the storm of the preceding night, I began to notice Dovekies—also known as Little Auks. Mostly I had only fleeting glimpses of them, as they rose momentarily above the surface of the heaving sea, to snatch a hurried breath before they again submerged to search for the small pelagic animals on which they subsisted. Although they are highly gregarious and swim in immense flocks, I detected them only at long intervals. Their dark upper plumage blended well with the dark, rough surface of the sea; and when occasionally they rose above it, their white underparts were confusingly similar to the white-capped crests of the waves. Once two Dovekies emerged at the same time and, keeping close together, flew rapidly low above the surface until lost to view in the distance, but this was most exceptional. As far as the Grand Banks, we voyaged through five hundred miles of cold and stormy sea where these hardy pelagic wanderers abounded. Although I could not watch them prepare for the night, as I did with the more aerial kittiwakes, I did not doubt that they slept, or took whatever repose they needed, on or in the ocean. They were too far out to return to land.

Dovekies belong to the auk family, all of whose members are pelagic, finding all their food in the oceans, seeking a solid surface only to reproduce and sometimes to molt. From the towering cliffs, stony talus slopes, high earthen banks, steep grassy slopes, sometimes even the lofty limb of a tree where they hatch, they go to sea, some when less than half-grown, with wing feathers developed only enough to help them flutter down to the water, others as downy chicks a day or two old. There they live, at first fed by their elders, until they are strong and agile enough to collect their own food and until, often after several years, the urge to reproduce sends them shoreward, where they may pass a season in often crowded breeding colonies before they become parents. Murres, guillemots, razorbills, murrelets, auklets, and puffins, as well as Dovekies—collectively called alcids—pass most of their lives in the cooler waters of northern seas, where undoubtedly they take whatever repose they require, although it is difficult to learn just how, or how much, they sleep. Marine birds tend to be active by night as well as by day.

No less tightly bound to the oceans than the alcids, and far more widely distributed, over southern and tropical seas as well as northern waters, are the four families of tubenoses of the order Procellariiformes, the albatrosses, shearwaters, petrels, storm-petrels, and diving petrels, including fulmars and prions. Like alcids, they seek solid ground only to breed or, in the case of birds just becoming adult, to prepare themselves for reproduction, passing a season or two with a mate at a breeding site, sometimes making a nest, without producing an egg. Some breed in immense colonies on remote or coastal isles, or in fewer numbers high on tropical mountains, in open nests or burrows that they dig for their single white egg. Some species undertake long migrations, traveling to far northern waters after nesting at high southern latitudes, or the reverse. Except when breeding, only on the water can they repose.

The three species of tropicbirds are also highly pelagic. After raising their single chick on small islands, they spread over the oceans, where they may spend several months, capturing flying and other fishes and squids by swooping over the water or plunging into it, and molting. Although I have found no definite statement, they can hardly rest otherwise than upon the surface of the sea.

Among flightless penguins are species that stay on land only to nest and molt and those that sleep ashore at all seasons. The former include the Adelie, Chinstrap, Rockhopper, Erect-crested, Snares Crested, and many others. Because of their prolonged nesting season and the slow development of their young, King Penguins are associated with their breeding colonies on islands in cold southern seas throughout the year; but between incubation spells, at which the sexes alternate, they may remain at sea for twelve to twenty-one days. While male Emperor Penguins incubate their single egg on snow and ice in the darkness and biting gales of the Antarctic winter, their mates remain for about two months at sea, eating well and accumulating reserves that will enable them to feed their chicks when they return to the colony about the time the eggs hatch. During the summer months, these largest of penguins live at sea, where they dive very deeply for food.

In a less rigorous climate, Galápagos Penguins sleep on the islands. Between five and seven o'clock in the morning, they go to spend the day foraging in the sea, not without landing for occasional short rests. Beginning at about four o'clock in the afternoon, and in numbers that increase until sunset, small groups of these penguins come ashore for the night. During the incubation period, one parent stays with the eggs while the other enters the water in the morning. Likewise, Yellow-eyed Penguins of New Zealand, who do not wander far from their nesting places, appear always to sleep on land. On any evening in winter, when these penguins do not breed, each may

Galápagos Penguins

be found at its own colony, where it will pass the night. These birds appear anxious to reach the land before it is quite dark, probably because they fear enemies in the water.

Although less continuously associated with the oceans than alcids, tubenoses, and penguins, certain birds that nest, and in the breeding season find all their food, far inland, after raising their families pass long months at sea, becoming for a time hardly less pelagic than murres, shearwaters, and albatrosses. After a short breeding season amid boreal forests or on the Arctic tundra, Red-necked, or Northern, Phalaropes and Red Phalaropes (known in Britain as Gray Phalaropes) migrate southward, the sometimes polyandrous females departing first; next the adult males, who incubate the eggs and lead the chicks; and finally the young of the year, when they are strong enough to undertake the long journey. During the northern winter, they spread widely over warmer oceans rich in plankton, many advancing far into the southern hemisphere. The third species of phalaropes, Wilson's, breeds in inland ponds and marshes of temperate North America and winters largely in South America, where it is frequent on the pampas of Argentina.

Few birds are as well fitted for a wholly aquatic life as the grebes, who not only eat, sleep, and court their partners on the water but likewise incubate their eggs in nests of vegetable materials that float on shallow inland waters, anchored to emergent plants. Unlike certain other birds of wet places which, as we shall see, build platforms to brood their young above the water, grebes carry their downy chicks upon their backs while they swim and dive, and often feed them there; or sometimes, as has been reported of the European Dabchick, or Little Grebe, the young return at intervals to rest on the nest until they are almost fully fledged. After raising their families, grebes that breed at northern latitudes where ponds freeze migrate southward and often seaward, flying overland by night, preferably by moonlight, and swimming through coastal waters. Several species of northern grebes, including the Eared, Horned, Red-necked, and Western, spend the winter months mainly on salt water in the coastal zone. Here they rest or sleep, pushing their bills into the plumage of their necks and pressing one or both feet against their flanks and covering them with their wings, an arrangement that helps to conserve heat.

Birds that breed on freshwater lakes and ponds but during the winter months forage and sleep on coastal salt water (as well as sometimes on inland fresh water) include the loons or divers. Although ducks are predominantly freshwater birds, some winter off the coasts. American Scoters and White-winged Scoters pass winter nights massed together in great "rafts" on sheltered waters off the southern coast of New England, far enough from land to feel secure from gunners. Common Eiders nest on seaside cliffs and small offshore islands at high latitudes around the Earth, and at all seasons forage for mussels and other marine invertebrates. In winter they frequent shallow coastal waters, often as far north as they can find openings in the ice. At night they rest on lonely islets, rocky ledges, drifting icepans, and perhaps also on the sea itself.

Some birds that find their food in the water must rest on dry land, rocks, or trees, a need which sets a limit to their wanderings over the oceans. This is especially true of cormorants, whose plumage soon becomes soaked, impeding their flight and obliging them to spread their wings to dry in the sunshine. Guanay Cormorants, who forage in vast numbers in the cool, nutrient-rich waters along the coast of Peru, return each evening throughout the year to sleep upon the same barren islets and rocks where at certain seasons they nest. Long ago, I enjoyed the rare privilege of visting a Peruvian guano island in the company of William Vogt, who was employed by an agency of the Peruvian government to study the guano-producing birds, and Herbert Hoover, Jr. The time was August, winter in the southern hemisphere, when none of the Guanayes was breeding, and they spent their days far off over the ocean. The afternoon, unlike most at this season, was clear. As the guano company's launch sped outward from the port of Callao, little waves sparkled in the bright sunshine, and more and more of the Andes spread before our view, summit beyond lofty summit, all brown and bare and craggy and forbidding.

Landing at the company's pier on the Island of San Lorenzo, we climbed several hundred feet up a steep, dusty slope, quite barren of vegetation, to the windward southern end of the ridge that runs the length of the island, where alone the Guanayes nested, the leeward slopes being too warm for them.

While we crossed the water in mid-afternoon, long, sinuous, black lines of Guanayes were already advancing toward their resting ground, their day's fishing done. The birds flew in single file, one close behind another, necks stretched forward in typical cormorant fashion. Crossing the island's crest, we saw where they were settling down to rest and sleep. They had only begun to come home, yet thousands were already on the ground. As we watched, more and more flew in from the sea, until the whole long slope was black with them. Their low grunts, multiplied ten thousand times, formed a continuous roar. They stood in close formation, the outermost in a straight line on the side toward their human visitors. As we advanced toward them, they walked deliberately away, with a loud pattering of hundreds of broadly webbed feet, until their compact ranks made an even semicircle around us, for they consistently maintained the same distance from us. As I moved slowly into the semicircle, it retreated ahead while it closed behind me, until I stood in the center of a living circle of crowded black necks and immaculate white breasts and bellies, while a thousand pairs of bright yellow eyes stared at me from bare red faces. Every bird kept its bill wide open and cheeks and throat vibrating, by this "gular flutter" increasing evaporation of water from the mouth and dissipating excess heat on this sunny afternoon.

It was now about four o'clock, and the sun was sinking toward a distant horizon. The Guanayes were returning from their day's fishing in increasing numbers, always swinging around to reach their resting ground from over the cliffs below it instead of flying over the crest of the island. Flock after flock was winging toward me, the serpentine files of birds appearing like a compactly massed formation when viewed directly from the front. The air above me was full of cormorants, banking their wings and dropping their feet to reduce their speed as they looked about for a little clear space where they could alight in the midst of the already crowded birds. I marveled that they could land in such small areas. One poor bird came down in the empty circle at whose center I stood and, perhaps distracted by the strange presence, fell upon its chin. Blue-and-white Swallows circled above the Guanayes massed upon the ground, snatching insects from the air. Far below, Brown Pelicans rested upon rocks or flew slowly above the waves. Kelp Gulls, beautiful in black and white, soared around looking for refuse. Piqueros, as Peruvian Boobies are called on the guano islands, plunged into the sea from high above it.

Very different from the Guanayes' way of sleeping was that of about two thousand Double-crested Cormorants who fished in San Francisco Bay in California, and at night roosted, like blackbirds on a telephone wire, on a power line stretched above the water. Standing erect, the big birds grasped a wire with both webbed feet and turned their heads to right or left to rest upon their backs, with the bill but not the whole head covered by one wing. In a strong breeze, they turned as far as the direction of their perch permitted to face into it. Less birds of the ocean than Guanayes, Double-crested Cormorants breed in the interior of the North American continent as well as along both of its coasts. I never cease to marvel at birds' ability to perch calmly on uninsulated electric wires charged with lethally high voltage. Nevertheless, they remain unharmed unless their wings simultaneously touch two of the wires, when they are instantaneously electrocuted.

I did not learn where the Piqueros of San Lorenzo rested, but probably, like other boobies, they did so on this or some other island, instead of on the water. On Raine Island, near the northern end of Australia's Great Barrier Reef, John Warham found about two thousand Brown Boobies without family ties resting on the beaches, in long lines two or three deep, all facing the sea. On a neighboring sandbank, more than a thousand Brown Boobies thronged the beaches to rest. Red-footed Boobies roosted in large companies on low, scrubby bushes, about 6 inches (15 centimeters) from the ground, or in dense grasses where they almost touched the ground. Often two or more slept in contact. When half awakened by a flashlight's beam, they would preen themselves or their neighbors with a nice impartiality, usually keeping their eyes tightly closed. In addition to these gatherings of hundreds of Red-footed Boobies, many solitary individuals, pairs, and trios slept on bare branches that were mostly sites of earlier nests. In Belize, Jared Verner found the off-duty members of pairs roosting at night on perches from 10 to 25 feet (3 to 8 meters) distant from the nests where their partners incubated.

Each pair had a single night roost, which might be shared with other pairs.

On islets beside Ascension Island in the tropical Atlantic Ocean, established pairs of Masked, or White, Boobies defended their nesting sites even outside the breeding season and tended to return to the colony at night. Adults not engaged in nest attendance as well as juveniles slept in "clubs" on the outskirts of the breeding colonies. Masked Boobies sometimes rest on the backs of turtles floating on the ocean, but apparently rarely, if ever, directly on the water. At night, the off-duty member of a pair of Brown Boobies stands beside its incubating partner. After their nestling grows too big to be brooded, they may pass the night on the ground with the young booby between them. For weeks or months after they fly, juvenile boobies return to the nest sites to be fed by their parents. In all these ways, the boobies of tropical seas differ from gannets of colder waters. Before they can fly, young gannets flutter from high, rocky ledges down to the ocean, where, abandoned by their parents, they live on their heavy deposits of fat until they can forage for themselves. Until they return to the breeding colony when a year or more of age but still too young to reproduce, they remain continuously at sea. Migratory gannets cannot return nightly to their breeding place to rest, as many boobies do.

Like cormorants, frigatebirds wear plumage that readily becomes soaked with water, preventing flight. To avoid wetting it, they neither swim nor dive, but catch flying fishes in the air, pluck food from the surface of the ocean, or harry other birds until they drop their food, which the piratic frigatebirds catch while it falls. Needless to say, these birds do not rest at sea but roost on bushes of the islands where they nest. However,they are sometimes seen so far from land that, it is surmised, they must sleep on the wing.

A bird which, on stronger but largely negative evidence, rests or sleeps in the air is the Sooty, or Wideawake, Tern. Its plumage so readily absorbs water that it could hardly survive an hour on the surface of the sea. Nevertheless, when not breeding, it is not known to alight on the islands where it nests, although its distinctive *wideawake* call may be heard from birds flying overhead through the hours of darkness, and it has not been found resting on land anywhere else, all of which points to the conclusion that for months together it remains continuously airborne, probably beating its wings slowly while it dozes.

It is improbable that any other tern sleeps in the air. At sea, and on wide rivers and lakes, terns, who dive for their food, rest on logs or other flotsam rather than swim in the water, as gulls more often do. Flocks of mixed species are often seen loafing or drowsing on sandbars, wide beaches, or other fairly safe places. On Raine Island, when they were not nesting, Brown Noddies were present all day, but their numbers increased in the afternoons, when they gathered in long lines on the beach, close to the water's edge. Some settled on the dunes to sunbathe, spreading wings and tails to the rays. When it was almost dark, many more noddies dashed in from the sea, skimming over the dunes and ridges in erratic flight, like nightjars. Gradually they settled down in dense packs on the island's central flats, cawing and cackling in chorus for a while before they slept. Late risers, the noddies lingered where they rested until several hours after sunrise. With the Brown Noddies slept a few White-capped Noddies. At Pandora Cay, ten miles from Raine Island, about five hundred Crested Terns slept on a sandspit. On Lancelin Island, about twenty Bridled Terns slept on low saltbush or on the ground amid vegetation. They were restless, calling at intervals through the night. Near them a party of Fairy Terns slept on the sandy beach. On Ascension Island, Black Noddies, when not breeding, slept on ledges of the cliffs where they had nested.

Except phalaropes, the oceanic birds mentioned in this chapter breed in monogamous pairs. Both sexes alternately incubate the egg or eggs and brood the young; both feed them. The intervals between changeovers during incubation, which determine how long each member of a pair remains continuously with the eggs, vary enormously from species to species, and even in the same species. Similarly, the intervals between feeding the single chick that many of these birds rear are most variable. These intervals are, in many instances, related to how the birds sleep, which in turn determines how far from their nests they can forage and how long they remain away. Very long sessions of incubation are taken by many tubenoses, who come to land only for breeding and rest or sleep at sea. While nesting on the Isle of Skokholm off southwestern Wales, Manx Shearwaters often forage in the Bay of Biscay, off the coasts of France and Spain, 500 and 600 miles (800 and 950 kilometers) from their eggs and chicks. Their spells of incubation range from three to ten days. Sooty Shearwaters may remain on their eggs as long as thirteen days. Sessions lasting two or three weeks are not unusual among the large albatrosses. Great-winged Petrels sit continuously for thirteen to more than twenty days. After fasting for these long intervals, the birds pass corresponding periods at sea, while their mates take charge of the nests. They may travel very far, seeking food, without touching land. The incubation sessions of storm-petrels tend to be shorter

—about four days for Leach's, two for Wilson's—yet they entail fasts that are surprisingly long for such small birds. At the other extreme, the Common Diving Petrel, which forages in coastal waters and apparently does not range far from its nesting burrows, incubates for about a day at a stretch.

The situation among penguins is especially informative. Yellow-eyed Penguins, who sleep ashore at all seasons, replace each other on the nest at intervals of one or two days; rarely one sits as long as five days. Similarly, the incubation spells of Galápagos Penguins, who rest on land every night, average about two days, with extremes of one and ten days. Very much longer sessions of incubation, often continuing for two or three weeks or even more, have been recorded of penguins that remain at sea when not breeding or molting, including the King, Adelie, Rockhopper, and crested penguins. Male Emperor Penguins remain on the ice, fasting, for the whole incubation period of about sixty-two days. Some penguins abstain from eating for intervals much longer than their maximum sessions of incubation, for they remain ashore for days or weeks before they lay their eggs.

The incubation spells of boobies, whose foraging range at sea is limited by the fact that they return nightly to sleep on land, are measured in hours rather than days: Brown Booby on Ascension Island, one to twenty-four hours, rarely longer; Masked Booby on the same island, thirteen to forty-eight hours; Red-footed Booby in Belize, about twenty-four hours; North Atlantic Gannet, mostly about twenty-four hours, with extremes of seven and one-half and thirty hours. Tropicbirds, who often nest on the same islets as boobies, take longer sessions on their single egg, often sitting from three to six or more days, while their partners remain at sea.

Most gulls and terns sleep on land. Herring Gulls sit by day for periods of about two to five hours; Common Terns for from a few minutes to about four hours; Black Terns rarely as long as an hour and a half; Black Noddies for about twenty-four hours. The Sooty Tern, whose habit of passing nights in the air permits it to range more widely, was found on Ascension Island to incubate for intervals of 89 to 157 hours, with a mean of 132 hours ($5\frac{1}{2}$ days). On other islands, such as the Dry Tortugas and Seychelles, Sooty Terns' incubation sessions may be much shorter, but longer than those of other terns. Cormorants, whose absorbent plumage does not permit them to remain long in the water, replace each other on the eggs at fairly short intervals: from one to three hours in the Double-crested Cormorant; about four times a day in the Shag. On the other hand, frigatebirds, who try not to soak their plumage but can remain airborne for much longer periods, often incubate continuously for days: up to fifteen in the Great Frigatebird; as long as six in the Lesser Frigatebird.

Intervals between nestlings' meals correspond roughly to the lengths of their parents' sessions of incubation and brooding; those that replace each other on eggs or young more frequently feed the young more frequently than those that sit unrelieved for very long intervals. The latter may feed their chick only at intervals of days, compensating for infrequent meals by making them very copious, sometimes regurgitating to the young bird food that weighs almost as much as the recipient itself.

It is often difficult to learn where and how oceanic birds rest or sleep, or how far they go for food between turns at incubation or while feeding their young; but knowledge of the lengths of their sessions on the nest or of the intervals bewteen their nestlings' meals may provide helpful indications.

References: Ashmole 1962; 1963; Bartholomew 1942; 1943; Bent 1925; Boersma 1977; Campbell and Lack, eds. 1985; Diamond 1975a; 1975b; Dorward 1962; Harris 1973; Imber 1976; Kendeigh 1952; Lockley 1942; Perry 1946; Rice and Kenyon 1962; Richdale 1943; 1951; Stonehouse 1953; 1960; 1962; Tuck 1960; Verner 1961; Warham 1961; 1974a; 1974b.

TWO

Birds of Marshes, Swamps, and Inland Waters

Marshes, shallow ponds, and lagoons are often rich in foods for birds but, when devoid of trees and shrubs, lack dry sites for sleeping, nesting, and brooding the young. Less mobile than oceanic birds, the feathered denizens of these watery habitats do not seek for their nests solid ground, which may be far from their foraging areas, but breed among the sources of their food. By diverse means, they have solved the problem of keeping themselves, their eggs, and their chicks sufficiently dry. Some incubate on floating nests of vegetable materials, which, sinking as they become sodden and heavy, must at intervals be built higher by placing fresh pieces on the top. Grebes, loons or divers, and swans of middle latitudes carry their downy young upon their backs, sometimes feeding them there. Pheasant-tailed Jacanas and probably other members of the family push their wings between their eggs and the damp floating vegetation upon which they incubate. Other marsh-dwellers build special platforms or nests for resting, courting, and brooding their young.

The large, black-and-white Magpie-Goose of northern Australia breeds in colonies in swamps where dense stands of sedges and grasses, especially spike-rush (*Eleocharis*) and wild rice (*Oryza*), grow above 2 or 3 feet (60–90 centimeters) of water. In January, when the swamp begins to flood and emergent vegetation to grow, the geese arrive and build stages. With their bills they grasp vertical stems that rise above their heads, bend them down until they can be grasped by a foot, then trample them into a platform above the water. Or, if the stems grow close together, a goose encircles with its neck a cluster of them and lowers them to its feet. Magpie-Geese breed in pairs or trios of a male and two females. The males initiate stage-building and perform the major share of the work. These stages are

Magpie-Geese building stage

used briefly for resting, preening, and courting, before the geese move on to build new platforms elsewhere. Rarely they return to reoccupy old ones.

Magpie-Geese begin to nest in March. As the time for laying approaches, stage-building passes into nest-building. Now the geese break off or pull up lengths of sedges or grasses and lay them upon the platforms of bent-over stems. The first of these nests are often incomplete and used only for resting or courtship. The finished nests are substantial floating structures, freshly built on the morning when the females start to lay, the two mates of bigamous males in the same nest. Single females lay from 5 to 12 eggs (average 8.6); two females from 7 to 12 (average 9.4), each producing fewer eggs than a single bird does. After twenty-four or twenty-five days of incubation, the goslings hatch, and spend a day in the nest before they follow their parents through the swamp. When only two or three days old, they perform all the movements of stage-building. During the ten weeks that they remain in the breeding-swamp before they can fly, they continually build stages, the only places where they can rest and sleep in the wide expanse of inundated land.

In addition to egg-nests floating in water and anchored to emergent plants, Common Moorhens, or Gallinules, build brood-platforms, or they brood their downy, precocial chicks on structures made by

American Coots or Muskrats. As daylight wanes, a parent, of undetermined sex, settles on a brood-platform and calls *kuk-kuk-kuk*. Approaching singly, the chicks peep as they climb the ramp that leads up to the platform, beg briefly, then push beneath the parent, who rises slightly to receive them. This is repeated until the whole brood of five to ten is snugly ensconced above the water.

In freshwater marshes of eastern and southwestern Australia, Dusky Moorhens live in cooperating groups of from two to seven individuals, with from one to three males for each of the females, who mate with all the males in their territorial group. All group members help to build nests above water, some of which are approached by ramps leading up to the rim. All the females lay in the same nest, each adding from five to eight eggs to the composite set, and all group members, of both sexes, incubate by turns for about three weeks. When, at the age of about three days, the chicks leave the egg-nest, they are led to nursery nests newly built for them, usually close above water deeper than that where egg-nests are situated. Here they are brooded by some of the adults by night, and by day in inclement weather, until they are four weeks old. After this, they sleep by themselves, first on low platforms and later on roosting platforms like those built by adults. Made of trampled vegetation, from about 20 inches to 6 feet (0.5 to 2 meters) above water in reed beds or trees, these platforms are loosely grouped, with 20 inches (50 centimeters) or more between them. On these the moorhens sleep singly, standing on both legs, with their heads tucked into the feathers of the back. Although they occupy the platforms repeatedly, they do not always use the same one every night.

Another member of the rail family, the Pukeko or Swamphen, inhabits marshes in New Zealand, where it lives in pairs or cooperating groups of from three to six individuals eight months or more of age. In this species, only males were seen to build nests, amid emergent stands of cattails or bog rushes (*Juncus*) or in tussocks of sedges. The egg-nests are cup-shaped structures of interlaced leaves, approached by a ramp and screened by overhead vegetation. The females of a group lay their eggs in the same nest, and all group members of both sexes take turns incubating them for about twenty-five days. After the eggs hatch, the adults build platforms or brood-nests above the water, to which, within three days of birth, the chicks are led, to be fed there by day and brooded at night. Sometimes egg-nests among the cattails also serve as brood-nests. Nests used for brooding are frequently abandoned, to be replaced by others nearer productive feeding areas.

Coots, unlike most rails, have lobed toes for better propulsion in water and are among the most aquatic members of the family. American Coots build three types of nests amid cattails or other emergent vegetation, close to open water. The first of these types is the display platform, coarsely built of cattail stems and leaves, seldom rising more than an inch above the water, and allowed to disintegrate after eggs have been laid. Until it falls to pieces, the nonincubating member of a pair may sleep upon it. The egg-nest is more carefully made, with a foundation of coarse stems, a lining of fine dry leaves, a rim rising from 4 to 6 inches (10 to 15 centimeters) above the water, and a ramp that permits the birds to enter from the water without tearing down the rim. Often several such structures are built before eggs are laid in one of them; the others may be used for sleeping at night. Each of three pairs of coots studied by Gordon W. Gullion in California built nine nests in a season, during which they laid two or more sets of eggs. In their zeal for building, coots remind one of wrens, who also construct nests for sleeping as well as for eggs.

The coots' four to ten buffy eggs, finely spotted all over with purplish brown, are incubated from twenty-three to twenty-five days, chiefly by the male parent, who covers them at night. Since the eggs are laid on consecutive days and incubation starts with the first, the downy black chicks may hatch over an interval of a week or more. While they are breaking out of their shells, the parents usually build a new nest, like the egg-nest but larger, up to 18 inches (46 centimeters) across and 8 inches (20 centimeters) high, most often without a rim. Or an egg-nest may be converted into a brood-nest. While the last eggs are hatching, the mother broods the older, more active chicks on the brood-nest, leaving the father to warm the remaining eggs and the belated chicks on the egg-nest. When the chicks are about two weeks old and are seldom, if ever, brooded by day, the young family may be divided at night, each parent covering some of them on a separate brood-nest. How long the young are brooded is not known; but one mother slept on a brood-nest with three of them about seven weeks old, possibly sharing the platform with these large juveniles instead of covering them. After this, these young coots slept on the nest by themselves, while their parents rested on nearby floating debris. In winter, when migratory coots settle on lakes, slow-flowing rivers, and marshes in compact flocks that may cover acres, they apparently sleep upon the water.

A number of other members of the rail family, including the Virginia, Sora, and Clapper rails, Spotless Crake, Black Coot, Red-knobbed Coot, and Purple Gallinule, build so-called "dummy nests" for brooding their chicks and sleeping. Sleeping in or on nests is not confined to rails of watery habitats. Thomas Gilliard found three

Gray-necked Wood-Rail on roosting platform

Forbes Rails sleeping in a nest of leaves and bark 11 feet (3.4 meters) up in a pandanus tree, 9,500 feet (2,900 meters) above sea level in New Guinea; and natives brought to Dillon Ripley pairs of Red Mountain Rails that they discovered sleeping in "little houses" in the high mountains of the same island.

Here in southern Costa Rica, the widespread Gray-necked Wood-Rail prefers tall thickets and light woods, especially near streams, but forages on firm land; I have never seen it in the water. It builds bulky open nests well above ground amid dense, concealing foliage, in situations that offer plenty of branches for roosting. Nevertheless, adults sometimes sleep upon platforms. I found one of these structures 6 feet (2 meters) up in a shrub beside a small, marshy opening amid dense, bushy growth. The rimless platform, composed of dead leaves, weed stalks, and other vegetable fragments, measured 10 by 12 inches (25 by 30 centimeters) across the top, and about 4 inches (10 centimeters) in thickness. Surmising that this was a wood-rail's sleeping platform, I stole up to it under cover of darkness, as silently as the tangled vegetation permitted. My flashlight revealed a grown wood-rail whose sleep I had interrupted, sitting beneath the canopy of vines, staring into the blinding rays with big red eyes and nervously twitching its short tail. After one good look, I turned off the light and crept away, leaving the rail on its platform. These rails, like many others, are sometimes active by night as well as by day. We hear their stirring *chirin-co-co-co* duet ringing out after nightfall, most often while the moon shines.

Unlike rails, ducks do not build nests for brooding their ducklings; nor do they, like grebes, loons, and some swans, commonly carry them on their backs. However, the ponds and prairie sloughs on which many kinds of ducks nest are often inhabited by Muskrats, whose massive lodges rise above shallow water. Blue-winged Teals, Redheads, Buffleheads, and other ducks have been found brooding ducklings and sleeping upon these convenient rodents' houses. The ducks also use matted floating or stranded water plants in the same way. Adult ducks and large juveniles sleep upon fresh as well as salt water, often massed in large "rafts." Of the Black Duck, Arthur Cleveland Bent wrote:

> When the swamps, ponds, and lakes of the interior are closed with ice the black ducks are driven to the seacoast to spend the winter. They linger in the lakes, even after they are partially frozen over, as long as an open water hole remains, resorting to the spring holes and open streams, visiting the grain fields and marshes or other places where they can find food and resting during the day in large flocks on the ice, where they sleep for hours while some of their number act as sentinels. On the coast their daily routine is to spend the day at sea or on large open bays and to fly into the marshes, meadows and mud flats to feed at night. At the first approach of daylight, long before the rosy tints of sunrise have painted the sky, black ducks may be seen, singly or in small scattered parties, winging their way out to sea, high in the air, their dark forms barely discernible against the first glow of daylight. At a safe distance from land they rest on the tranquil bosom of the sea or sleep with their bills tucked under their scapulars. It must be half-conscious sleep, or perhaps their feet work automatically, for they never seem to drift much. When the open sea is too rough their resting places are in the lee of ledges in little coves or in bays. Often times they rest and sleep in large numbers on drifting ice, on sand bars or even on unfrequented beaches. (Bent 1923, pp. 62–63)

Other ducks that rest and sleep on the sea by day, but at night fly in large flocks to forage in shallow inland waters are Canvasbacks and Northern Pintails. Red-breasted Mergansers do just the reverse, fishing in the daytime in coves, harbors, tidal estuaries among salt marshes, and shallow water along the beaches, and flying at sunset to sleep with greater safety on the ocean. Large flocks of Canada Geese sleep on ponds or lakes and forage by day on grain fields and pastures.

Sandhill Cranes rest, often in large groups, in ponds and lakes, in water less than 9 inches (23 centimeters) deep, or on mudbanks, sandbars, or other open places which permit a few watchful sentinels to detect approaching enemies while they are still distant. Exceptional in the crane family is the Wattled Crane of Africa, which roosts in trees. The curious Hammerheads of Africa roost in trees or sleep in their massive, well-enclosed nests, built of sticks and mud, with a long entrance tunnel. Often they lodge in an old nest.

Other long-legged birds of marshes, swamps, and inland waters, including herons, egrets, ibises, spoonbills, and storks, which nest, often in great colonies, in trees or bushes, roost in similar situations. Although American Wood Storks roost in trees, Maguari Storks of South America sleep on the ground. Night-herons, Boat-billed Herons, and some other long-legged birds rest by day and forage at night. Early in the year, a flock of fifteen Little Blue Herons, four slate-blue adults and eleven pure white young birds, roosted above the Río Pacuar, a swift-flowing inland stream of clear water in southern Costa Rica. They rested on densely foliaged lower branches of Riverwood trees stretching above the narrow, rocky channel, wholly exposed, the white young herons conspicuous against a background of dusky verdure. Before sunrise, while the valley was veiled in mist, they would fly upstream, closely following every curve, to pass the day hunting insects amid the grass, dry and brown at this season, in pastures on flat bottom lands and lower parts of the surrounding slopes. In the cool of evening, when the valley was in shadow and only the high cordillera in the north was aglow in the rays of the sinking Sun, they would return downstream, flying in close formation above the crowns of the riverside trees, to their roost.

A roost of Scarlet Ibises and egrets so spectacular that it draws crowds of tourists, including those with little interest in birds, is situated in the Caroní Swamp on the western side of the island of Trinidad. In late afternoon, our party embarked in a large, flat-bottomed boat, with an outboard motor that propelled us along a maze of natural and artificial channels through a forest of tall Red Mangroves. Emerging from this swampy woodland, we joined a number of similar, tourist-laden boats lined up in view of several small islands surrounded by wide expanses of open water. To these densely wooded islets the big birds streamed, flying singly or in loose flocks, until the dark foliage of the mangroves glowed with the intense scarlet of thousands of ibises and gleamed with smaller numbers of

Snowy Egrets and Cattle Egrets, with fewer Great Egrets. Although scarlet and white were intermingled in the crowded roosts, the ibises were concentrated near the tops of the trees, while the egrets tended to perch nearer the water. As the magnificent spectacle faded in the waning light, the boats full of onlookers curved widely around the roosts, never approaching close enough to disturb the birds, and followed the channels through the *manglar* back to the landing where the cars waited.

Unlike ibises, herons, and egrets, Sungrebes sleep alone. On many April evenings, a solitary bird, waving its head forward and backward like a walking chicken or pigeon, swam slowly and silently downstream, close to the bank of a smoothly flowing river in the lowlands of Costa Rica. At intervals the Sungrebe plucked small objects, probably insects, from the lush foliage overhanging the shore and from decaying logs. Then it bathed, dipping its foreparts into the stream and splashing water over its back with its wings. After wetting its upper parts, it rose to a branch or stranded log, to preen and arrange its plumage for many minutes. This might be followed by a second bath and more preening. Finally, while much daylight remained, it flew up to a long horizontal branch projecting from the bank, about 4 feet (1.2 meters) above the water. Here, beneath abundant sheltering foliage, it roosted until returning daylight was fairly bright, when it jumped down into the stream and swam upward along the bank, picking its breakfast from overhanging vegetation. The Sungrebe continued to sleep on the same branch until it was submerged by rising water in May. Sungrebes build small, open nests in situations similar to those where they roost.

References: Bennett 1938; Bent 1923; 1926; Craig 1980; Davies 1962; Erskine 1972; Frederickson 1971; Frith and Davies 1961; Garnett 1978; Gilliard 1958; Grey 1927; Gullion 1954; Littlefield 1986; Low 1945; Ripley 1942; Scott et al. 1972; Tennessee Ornithological Society 1943; Terrill 1943; B. T. Thomas 1986.

THREE

Ground-nesting Birds

lthough birds who rest upon the oceans perforce nest on solid ground or rock, inland birds tend to sleep and nest in similar places, tree-nesters in trees, ground-nesting birds upon the ground. This rule is not without exceptions, especially among ground-nesting birds, many of whom seek safer roosts in trees, although birds that nest above the ground rarely sleep upon it.

Great flightless birds, unable to perch, the Ostrich, Greater Rhea, and Emu, necessarily sleep upon the ground, like horses, cows, and other ungulates. Ostriches and rheas rest with their long necks stretched out on the ground in front of them, their legs extending behind, or doubled beneath their heavy bodies. In New Zealand, equally flightless Kiwis sleep by day in intricate subterranean burrows, to sally forth under cover of darkness and seek food in the earth with their keen sense of smell.

The tinamous of forests and thickets of the tropical American mainland and the temperate zone of South America regularly nest upon the ground, but their ways of sleeping are diverse. The Little Tinamous, Variegated Tinamous, and other species of *Crypturellus* rest on the ground amid low, sheltering vegetation. Douglas A. Lancaster described how Brushland Tinamous sleep in Argentina. When preparing to rest, a tinamou digs a shallow depression in the soil by scraping it backward with its feet, moving alternately. At the same time, it turns around, pushing the earth in all directions, until the hollow becomes about an inch (2.5 centimeters) deep. Then the bird settles in its depression and slides its body from side to side, as though adjusting eggs beneath itself. The tinamou may use its bill to draw earth and leaves against its body and become more snugly ensconced. Usually these tinamous make their hollows beside a tree or tussock of grass, where they are partly screened. The same sleeping

Great Tinamou roosting

site may be occupied repeatedly, or the bird may repose in a different depression each night. Seven Brushland Tinamous in an aviary slept a foot or two (30–60 centimeters) apart. On hot afternoons, some of these birds dug similar hollows for resting. They also took brief naps in the middle of the day while standing on one leg or two.

Although Great Tinamous incubate their lovely, glossy, turquoise-blue eggs upon the ground, often between buttresses of a great tree of tropical rain forest, they sleep well above it. William Beebe described how, in the land that was then called British Guiana, a Great Tinamou "stepped past with quick, dainty strides and half leaped, half fluttered awkwardly up to the base of a leaning tree, and with wildly balancing wings, made its way forty or fifty feet still higher to a large horizontal branch. Here without hesitation, backed close against the trunk, the bird squatted, and facing lengthwise of the branch, rested on its tarsi, which were applied closely to the rough, mossy bark"

(Beebe, Hartley, and Howes 1917, p. 258). In such an attitude, the roughly scaled posterior surface of the tarsus (lower leg) would help to keep the heavy bird from slipping off the branch. However, the Great Tinamou that I watched go to roost one evening in a Costa Rican forest flew directly up to a horizontal branch a few inches thick and about 20 feet (6 meters) above the ground, where it remained perching like any other bird until I could hardly see it without a flashlight, which did not disturb the bird. One afternoon, I saw a Great Tinamou fly in the same way to a much higher branch amid the forest, but it did not remain until evening. Sometimes in the night we hear the Great Tinamou's mellow organ notes floating out of the neighboring forest, from a wakeful or a dreaming bird.

Turning now to the pheasant family, we find that Northern Bobwhites commonly sleep on the ground in some sheltered spot, beneath an evergreen tree, thick bush, or dense tangle of brush or briars. Members of a covey arrange themselves in a compact circle, tails together, heads outward, so that collectively they can see in all directions. The first two bobwhites to arrive at the spot chosen for sleeping stand pressed together, facing the same way. Others attach themselves closely to this nucleus, forming an arc that, with later additions, grows into a closed circle. Those that arrive tardily forcibly push themselves between the massed bodies from the outside, or, standing upon the closely appressed birds, wedge themselves downward between two of them, all with much ruffling of plumage. If undisturbed, the quails may occupy the same spot for many nights, as revealed by a growing accumulation of droppings. If awakened by some prowling animal, the circle "explodes," its components flying to all sides with little danger of colliding with each other, and perhaps alarming the prowler by the suddenness of their dispersal. Day-old chicks already rest in circular formation, like adults.

Bobwhites do not invariably sleep on the ground in the compact ring that conserves their body's heat as well as affording a ready escape from predators. Occasionally they roost in a tangle of vines or a bushy tree, on a fence rail, on logs, or in a chicken house or barn. In winter they may dive into soft snow, or huddle together in a sheltered spot while falling snow softly blankets them. Gray, or Hungarian, Partridges, introduced into North America from Europe, also gather in a compact circle with outwardly directed heads to pass the night, or they sometimes roost on the limb of a large tree. In western North America, Mountain Quails and Scaled Quails sleep on the ground, but California Quails roost in bushes, low in trees with dense foliage, or, in treeless deserts, amid clumps of cacti.

Little is known of the habits of tropical American wood-quails, of

Northern Bobwhites preparing to sleep

which at least some species nest upon the ground. One evening in the mountains of Costa Rica, I watched a pair of Spotted Wood-Quails fly into a tree beside a coffee plantation, then rise from branch to branch until lost to view in the high crown.

Grouse, including the Blue, Spruce, and Ruffed, roost in trees, preferably densely branched conifers, or, in winter, in snowdrifts on the ground. Sage Grouse, Sharp-tailed Grouse, Greater Prairie Chickens, Rock Ptarmigans, White-tailed Ptarmigans, and Willow Ptarmigans all sleep on the ground or amid snow. Sometimes Willow Ptarmigans dive in swift flight deep into softly drifted snow, leaving never a footprint to betray their presence. In the morning, if all goes well, they burst flying from their cold chambers, again making no trail. At other times, they simply squat in shallow depressions in the snow. On Southhampton Island in the Arctic Ocean, seven slept for a night in as many consecutive footprints made by the artist George M. Sutton as he trod over snowy ground. Alternatively, Willow Ptarmigans tunnel into the snow, making it fly as they kick backward until they have dug themselves into the roofed chamber where they sleep. In shallower snow, Rock Ptarmigans dig down from the surface, kicking behind them not only snow but fragments of plants from beneath it, which, lying darkly on the white surface, may be more revealing than the white birds themselves, whose upper parts rise above the shallow depression in which they rest.

When the thermometer falls many degrees below the freezing point, small passerines take advantage of snow's insulation. On the Arctic tundra where they breed in crevices amid rocks, Snow Buntings sleep in depressions in rough, eroded ground, in shallow niches in stony outcrops, tussocks of grass, or whatever offers them a little protection from chilling winds. They also take shelter in deep pits in snowbanks, in groups but, apparently, not in contact with each other. While wintering in New England, Snow Buntings burrowed into the leeward sides of snowdrifts, not only to sleep by night but also to conserve heat on bleak wintry days, when they emerged only long enough to seek food in piles of chaff. In the Irkutsk region of Siberia, Willow Tits burrow under snow and sleep in abandoned rodents' burrows. An individual tit may retain the same hole throughout the long winter.

Snow appears to make a cold bedroom, but when soft and uncompacted its numerous air spaces make it a poor conductor of heat. The air surrounding a bird resting in a cavity in a snowdrift may be many degrees warmer than that outside. Often the sleeping bird preserves the same orientation through the night, with the result that the moisture in its breath congeals into a wall of ice in front of it; to leave, it must turn to the side. Opposed to the advantage of conserving vital heat by sleeping beneath snow is the danger that sleet and cold rain may cover the surface with a hard crust which the birds cannot break. Unable to escape, they may perish from cold and hunger in their icy prison. Moreover, prowling foxes or other mammalian predators may detect with keen noses the birds' presence beneath the snow.

Ring-necked Pheasants, who nest on the ground, may sleep either on the ground or in trees. Other ground-nesting members of this family, including the Common Peafowl, Helmeted Guinea Fowl, Red Jungle Fowl and its descendants, domestic chickens, roost in trees. Wild Turkeys also sleep in trees, often above water.

In dry, semi-open country, Double-striped Thick-knees are active by night, and by day rest in small groups in the shade of a thicket, often sitting upon their legs. Killdeers rest on open fields, pastures, sand, or gravel, either sitting or standing, often on one foot. Light sleepers, they are often heard in the night. In their winter range in tropical America, their plaintive cries often mingle with the softer notes of the nocturnal Common Pauraque. Easily aroused, when startled they take wing and circle around in the darkness. When high tide covers much of the beaches where they forage, Semipalmated Plovers, like many other shorebirds, sleep above reach of the waves. Here a number huddle together, some standing, others squatting with breasts resting on the sand, often in the lee of driftwood. Some sleep with heads turned to one side and bills thrust into the feathers of their backs in the usual manner of birds, others with their heads between their shoulders. Nearly always some members of the flock are awake and alert; others open their eyes from time to time.

In the sandpiper family, the Knot, Sanderling, Semipalmated Sandpiper, Hudsonian Curlew, Long-billed Curlew, and many others sleep on the beaches, principally while high tides flow over the most productive foraging areas lower on the shores. Since the activities of birds that find their food between tide levels are governed by tides that rise and fall twice in nearly twenty-five hours, they often rest for part of the daytime and forage for part of the night. Spotted Sandpipers, who frequent narrow inland streams as well as beaches, have been seen sleeping in large numbers on the tar-and-gravel surface of an airfield, in April, when they were apparently migrating northward. In Trinidad, in the same month, Michael Gochfeld discovered twenty-nine loosely associated Spotted Sandpipers roosting after sunset on prop roots and dead branches at the edge of a mangrove swamp. Resting from 6 inches to 5 feet (15 to 150 centimeters) above the water, some were only 6 inches apart. They also slept on mudflats and on a boat dock. Spotted Sandpipers are rather exceptional in their family in their ability to perch; but Wood Sandpipers, Willets, Greenshanks, Redshanks, and Upland Sandpipers (formerly Plovers) also alight in trees. Whether they sleep there I do not know.

Woodcocks, both the American and the European species, rest through the day on the ground in the deep shade of trees and bushes, to emerge in the twilight and seek food in meadows and wet fields rich in earthworms, which they detect by their acute sense of smell. By day woodcocks seldom fly unless disturbed, when they flutter through the branches in a zigzag course.

When not snuggled down upon the ground, shorebirds often stand upon one leg, with the other drawn up into their ventral plumage. Why do so many birds of diverse families spend long intervals in this posture that hardly appears restful? Unless the bird's center of gravity is exactly above a perpendicular leg with a precision probably difficult to maintain, some muscular exertion must be needed to prevent falling. Probably birds rest in this fashion to conserve heat. Our often cold hands, feet, ears, and noses remind us that the exposed extremities of a body at normal temperature are great dissipators of vital heat. Not only birds of cold climates adopt the

Sanderling sleeping on one leg

one-legged posture. One night I found two migratory Tennessee Warblers, not the most sociable of birds, sleeping in a cashew tree in our garden, pressed against one another, each resting upon its outer leg, with the inside leg drawn up into its plumage. The tropical night was only comfortably cool.

Although pigeons and doves usually nest and sleep in trees or shrubbery, exceptions are known. The White-bellied Plumed Pigeon of central and northern Australia lays its two eggs in an often unlined scrape in the ground. It sleeps in a similar depression in bare soil or stony ground on gentle slopes or near the crest of a hill, on the lee side of a small shrub, large rock, or clump of spinifex grass, never inside a shrub or under overhanging branches, which might impede escape from a nocturnal predator. In California, tree-nesting Mourning Doves departed from their usual habit of roosting in trees to sleep on the ground, most often in pastures, amid the stubble of hay fields, or other treeless expanses. On a night in August, forty-two of these doves were found resting in 60 acres (24 hectares) of open pasture. In winter, Mourning Doves often pass the night along unfrequented unpaved roads.

Among raptors, Short-eared Owls and Northern Harriers, known also as Marsh Hawks, nest on the ground, where they also sleep, the latter sometimes in loose groups of up to thirty individuals, in years when the mice that they eat are abundant. In Arabia, Northern Harriers rested on the ground with more numerous Montague's Harriers in concentrations of over two hundred birds. Year after year, I have watched Swainson's Hawks migrating along the Pacific slope of Costa Rica on their way from South America, where they winter, to the western United States, where they nest. Alternately spiraling high on thermal updrafts, then gliding forward on set, slightly closed wings, with loss of altitude, they cover long distances with little muscular exertion, seldom or never eating along the way. One afternoon in mid-April, as the Sun sank low and the drafts that bear them upward died away, I watched a great flock glide over the forested crest of a high ridge and slowly, sedately circle down to a steeply inclined pasture. The majority found perches on trees standing isolated in the closely cropped pasture, at the forest's edge, or beside the river at the foot of the slope. Others alighted on stumps and logs. Many more settled on the ground, where they remained until the rising Sun heated the east-facing slopes and created the thermals that they needed to continue on their journey. After the last of them had vanished into the blue, I searched the pasture for traces of their nightlong presence, finding never a dropping, a casting from the mouth, or a molted feather. The great host of Swainson's Hawks had departed as silently as it had come, leaving not the slightest memento of its passage.

Nightjars, frequently miscalled goatsuckers, commonly nest on the ground or some substitute for it, such as the flat, graveled roof of a building. Some, including nighthawks, roost by day in trees, resting lengthwise along thick or slender branches, instead of crosswise in the manner of perching birds. Whip-poor-wills and Chuck-will's-widows rest by day either on the ground or squatting lengthwise on low branches. Others, such as the Blackish Nightjar of northern and central South America and the Common Pauraque, far more widespread over tropical and subtropical America, nearly always rest upon the ground. The pauraque dozes throughout the hours of daylight in the shade of some second-growth thicket, or in a plantation of bananas or coffee, often near the edge of a clear space, where its coat of mottled blacks, browns, and buffs blends so well with the brown fallen leaves that a person might step on the bird without seeing it. Fortunately for the pauraque, it does not sleep too soundly to be alert to approaching footsteps. Aroused from its repose, it rises lightly and circles around in silent flight, revealing on wings and tail prominent patches of white, which vanish when it settles on the ground and folds its limbs, often at no great distance from the intruder and in full view. Rarely, when disturbed, the pauraque alights on a low branch with its body parallel to it.

How much the nocturnal pauraque actually sleeps by day is difficult to learn. In many hours spent in a blind watching by day a nest where the male did most of the diurnal incubation, I was impressed by the continual movements of his eyelids. Rarely he opened them fully or closed them completely for more than a few seconds together—half a minute was the longest interval of apparently full sleep that I noticed. Both lids moved up and down over the bird's large, dark brown eyes. At one moment they were closed except for a narrow gap over the posterior half; next they were everywhere tightly pressed together. A moment later they separated until a narrow slit extended across the eye's full width. Then they slowly drew together again. I have watched the same restless movements of the eyelids of the Common Potoo incubating or brooding its chick by day on an exposed limb of a tree. Even with eyes nearly closed, as though dozing, these nocturnal birds see well enough what is happening around them.

Some of the many passerine birds that build their nests on or near the ground have been reported to sleep upon it. Eurasian Skylarks pass their nights on the ground in open meadows. When nights are cold, Horned Larks of deserts and high mountains of the western United States prepare for them by digging with their bills holes in which to sleep. Each depression in the soil is just large enough to contain a single lark, with its back almost level with the surface of the bare surrounding ground. In soft soil a lark apparently digs a new hole every evening, but where the ground is hard it occupies the same one repeatedly. These little hollows are often situated where plants shield them from the prevailing wind. Here the larks sleep more warmly than on an exposed perch. Eastern Meadowlarks pass the night scattered singly on the ground in grassy fields. Savannah Sparrows sleep in small, compact groups in similar fields. White-throated Sparrows, Rufous-sided Towhees, and Dark-eyed Juncos repose in loose flocks on the ground in woods, where many leaves have collected in low undergrowth of briars and canes. Other juncos have roosted from 3 to 8 feet (0.9 to 2.4 meters) up in dense spruce trees. On cold winter nights, juncos sometimes snuggle down in an old nest low in a hedge of evergreens.

Sleeping on the ground, exposed to prowling reptiles and mammals, and to attacks by owls unless overhead cover is dense, appears to be particularly perilous. It is practiced chiefly by birds that forage or nest on the ground, some of which are poorly adapted for perching, because they are so heavy, or because their feet are ill-fitted for grasping; often the hindtoe is rudimentary. Many of the birds that rest upon the ground are more active at night, and more alert, than are passerines, which appear to sleep more soundly, only a small minority of them upon the ground. Equally important, on the ground birds often sleep in small or large groups, sometimes arranged in a manner that facilitates rapid escape, like Northern Bobwhites and Gray Partridges. A member of the group that happens to be awake warns its companions of approaching danger. Despite the perils of resting on the ground, birds that lack more adequate shelter do well to sleep upon it, for at ground level the velocity of chilling winds is much less than at a few yards above it.

References: Back, Barrington, and McAdoo 1987; Bagg 1943; Beebe, Hartley, and Howes 1917; Bent 1927; 1929; 1932; 1937; 1938; 1940; Bent et al. 1968; Cowan 1952; Gochfeld 1971; Goodwin 1967; Ingels and Ribot 1983; Lancaster 1964; MacDonald 1970; Parmelee 1968; Skutch 1945a; 1972; Trost 1972; Walkinshaw 1953; Weller, Adams, and Rose 1955; Wetmore 1945; Zonov 1967.

FOUR

Roosting on Plants and Buildings

Birds that nest above the ground nearly always roost above it, on trees, shrubs, tall grasses, cliffs, or artificial structures. With this broad similarity, their ways of sleeping are most diverse. They may rest alone, in pairs or families, in contact with each other or more or less separate, in large aggregations of one species or mixed gatherings of several species. Many birds prefer to sleep in an enclosed space, which may be a cranny in a tree, building, or cliff that they find waiting for them, a hole that they have carved in a tree, or a nest that they have built. In this chapter we shall look at birds who sleep in the more exposed situations, including open ledges on buildings. In the following chapters we shall consider those that provide roofs for themselves.

Solitary Sleepers

Unless birds of humid tropical forests lodge in holes or nests that can be located in the daytime, it is difficult to learn where they sleep amid exuberant concealing vegetation. Among the few forest birds that I have found in the night was a Gray-chested Dove, picked out by my flashlight while I prowled through lowland forest in Panama in search of sleeping birds. The dove rested upon a slender horizontal branch, about 10 feet (3 meters) above the ground, in a fairly clear space amid tangled growth. Its head was exposed, with bill pointing forward, as pigeons habitually sleep, whether on their nests or while roosting. I could find no other dove near it. After waking, it remained calmly on its perch in the beam of light, while I scrutinized it from all sides to make sure of its identity. Then I switched off the electric torch and stole away through the darkness. This was a memorable encounter, for I have found no other pigeon roosting in the undergrowth of tropical forest.

Although I have studied trogons for years, only once have I found one roosting. For nearly a month in April and May, amid the Caribbean rain forests of Costa Rica, a solitary male Slaty-tailed, or Massena, Trogon slept nightly at the forest's edge, beside an abandoned banana plantation choked with weeds. His roost was a densely foliaged branch leaning far out over the clearing, at a height of about 15 feet (4.5 meters). Soon after sunset he would appear in the top of a tall tree above the place where he slept. After resting for a while, he flew down to a lower branch and paused there, too. Then the brilliant trogon dropped abruptly down to the small tree where he roosted. He perched for a few minutes on a rather exposed branch before he moved into the clustered foliage where he passed the night, always alone. His time for retiring was a few minutes before six o'clock.

Although antbirds are widespread and abundant in the woodlands of tropical America and their nests have often been found and studied, little is known about their sleeping. The only published record that I know is by Allen M. Young, who discovered a Spotted Antbird roosting in the same forest where I found the trogon's roost. On most nights from early February to late May, the male antbird was present in the same tree in the forest undergrowth, and nearly always on the same branch, about 10 feet (3 meters) up. His time for retiring, from a few minutes before to a few minutes after six o'clock, was much the same as the Slaty-tailed Trogon's. Although Spotted Antbirds live in pairs throughout the year, no other individual could be found sleeping near the solitary male.

Another solitary sleeper was a Long-billed Starthroat. Its choice of a roost, no less than its manner of preparing for the night, were surprising. As daylight faded, the hummingbird alighted upon an exposed, slender twig at the very top of a dying guava tree in an abandoned pasture, with nothing but the sky above and only a few small leaves around it; it was visible from all sides. The site was even more exposed than that in a fork of a dead tree which starthroats habitually choose for their nests. After alighting on the twig, the hummingbird twitched its head rapidly from side to side. At first pronounced, these head movements gradually decreased in amplitude until they ceased entirely. On an evening when the hummingbird arrived at its sleeping place at 5:52 P.M., they continued for three or four minutes. On a darkly overcast evening, when the starthroat arrived at 5:36, they persisted for a quarter of an hour. When finally they died away, the hummingbird rested with head exposed, bill pointing forward and slightly upward, as I have always found hummingbirds sleeping, whether on their nests or while perching.

Hummingbird sleeping on twig

After dark, the starthroat's eyeshine in a flashlight's beam was surprisingly bright, like a brilliant point of light. Roosting on a nearly leafless slender twig at the top of a tree, the bird appeared to enjoy a maximum of security from terrestrial enemies at the price of full exposure to aerial predators such as owls, which were not numerous in this locality. A plainly colored Scaly-breasted Hummingbird that roosted in a tree beside our house likewise twitched its head from side to side after alighting upon its exposed perch. Both of these birds were quite alone.

Still another solitary sleeper was a speckled female Turquoise Cotinga, who for a few nights in May roosted on a slender horizontal branch, 40 or 50 feet (12–15 meters) up in a tall tree that shaded a small coffee plantation, a short flight from the forest where cotingas frequent the upper levels. Here she was screened above by a dense canopy of foliage but clearly visible from the ground. She went to rest at about half-past five o'clock and departed before sunrise. Equally alone was a wintering male Yellow Warbler who, from November to February, slept in a shrub of *Hibiscus mutabilis* beside the thatched cabin I then occupied in the mountains of southern Costa Rica. He perched upon a slender petiole, shielded above by the broad blades of higher leaves, but he was exposed on all sides and easily visible from the ground. A male Summer Tanager slept in orange trees behind our house, his red form conspicuous amid the dark foliage in a flashlight's beam. Since both Yellow Warblers and Summer Tanagers defend individual territories while they pass the winter months in the tropics, it is not surprising that they roost in solitude.

The sites chosen for roosting by many birds, screened above but fully exposed below, suggest that terrestrial predators are no great threat to them while they sleep. An approaching reptile or mammal would shake the branch and wake the sleeper, who would find it advantageous to drop below the perch and fly away with little danger of colliding with obstructing twigs.

Sleeping in Pairs

Continuously mated birds who sleep in scattered trees in gardens and clearings often perch near their partners. This arrangement is frequent among American flycatchers, including the Tropical Kingbird, Gray-capped Flycatcher, Boat-billed Flycatcher, and Yellow-bellied Elaenia. All these flycatchers often roost upon slender branches at no great height, canopied above by all the tree's foliage but exposed below. If the flashlight picks out the bright yellow breast of one of these birds, one can often find its consort sleeping from a few feet to a few yards away. One evening in September, I discovered a pair of Masked Tityras perching quietly in the top of a Burío tree beside our house, on petioles of the large, cordate leaves, about 40 feet (12 meters) up and a foot (30 centimeters) apart. Above them was a canopy of foliage, but their white underparts were plainly visible from the ground. Here they remained until the following morning, but in the evening of this day I looked fruitlessly for them. A year later, almost to the day, I found a lone female Masked Tityra roosting in the same Burío tree, where she slept only two nights after I discovered her. After going early to rest, while the Sun still shone, and becoming active late in the morning, these tityras flew from their isolated treetop directly to the upper levels of the neighboring forest where they foraged.

Tanagers often roost amid the abundant foliage of orange trees, whose long, stout thorns, projecting horizontally from the main trunk and upright branches, are convenient perches. Paired Golden-masked Tanagers and Blue-gray Tanagers usually sleep from a few inches to a few feet apart on thorns or twigs, sometimes with their young nearby. At times a large orange tree shelters several pairs of these tanagers, with Palm Tanagers, Bay-headed Tanagers, Buff-throated Saltators, flycatchers, a single Summer Tanager, and other birds. In the evening, they dart so rapidly into the screening foliage from all sides that it is impossible to count them. They sleep with heads snuggled into the plumage of their shoulders, and at dawn they shoot outward as swiftly as they came.

Sleeping in Contact

Some birds sleep pressed together on their perches. Fledglings, accustomed to intimate contacts in the nests that they have just departed, often perch close together even if, when older, they will roost separately. This is true of a number of American flycatchers, including the Vermilion-crowned, or Social, Flycatcher, the Tropical Kingbird, and the Common, or Black-fronted, Tody-Flycatcher, whose juveniles continue to line up wing to wing until well grown, sometimes with a parent pressed against them. Once I found three well-grown Rusty-margined Flycatchers perching in a compact row in a shrub above a rivulet, their yellow breasts all turned the same way. At the end of the day on which they first flew from their burrow in a bank, three Southern Rough-winged Swallows slept through a drizzly night pressed close together on a thin twig at the top of a dead tree, in a situation no less exposed than that chosen by the Long-billed Starthroat. Long after nightfall, I could detect them through my binocular as a dark thickening on the twig, silhouetted against an overcast sky. They slept in this hazardous situation, exposed to attack by owls, for only a single night.

Fledgling Wren-Tits in California rest by day, and probably sleep by night, in close contact. When preparing to sleep, members of an adult pair of Wren-Tits press close together and arrange their feathers with their bills until those of the two partners interlock, forming one downy mass with no visible line of separation. Into this mass the inside legs of the two birds are drawn and concealed. The outside legs incline outward, bracing the sleeping partners against each other. If several Wren-Tits are kept in a cage, they roost in one compact row. Even when roosting in flocks, mated Golden-fronted White-eyes of southern Africa perch in contact. Parents with fledglings huddle close to them, one on each side. Mated Gray-bellied White-eyes and Green-bellied White-eyes also roost in contact, and their fledglings sit bunched together by day.

Among pigeons, too, parents roost in contact with their flying young, as did a pair of White-tipped Doves with two almost full-grown juveniles. On successive nights the four doves arranged themselves in different orders on a branch of an orange tree. Sometimes three slept in close contact, with the fourth an inch or two from the trio. On other nights they rested two and two. They might all face the same way or adjoining individuals in opposite directions, with their heads exposed in typical pigeon fashion. Parent Inca Doves sleep as close to their juveniles as White-tipped Doves do. Even adult Inca Doves roost in intimate contact, in one instance with seven pressed together in a row and three more resting on their backs. At times a parent dove extends a protecting wing over the juvenile at its side, as has been reported of the Emerald Dove of India. As soon as they can climb up to a perch, downy flightless chicks of the Great Curassow of tropical America are brooded there beneath their mother's wings, one on each side—a charming sight. Brooding on the perch is practiced also by Great Argus Pheasants and Red Jungle-Fowls. Domestic hens, placed on a perch before they have ceased to brood, also shelter their half-grown chicks beneath their wings through the night.

Sleeping in contact is widespread among birds who breed communally, two or more pairs sharing a nest, or cooperatively, with helpers aiding a single breeding pair to raise their young. These highly social birds may huddle together in dormitory nests (at which we shall look in later chapters), roost in a compact row, or, less frequently, cling together in a ball. Among those that sleep on a perch, with up to a dozen pressed together in a row, are the black anis of tropical and subtropical America, both the Groove-billed and the Smooth-billed. Twelve to twenty of the related Guira Cuckoos of southern South America roost pressed side by side, with some on the backs of others in two or three tiers. Jungle Babblers of India line up on a perch, all facing the same way, with the breeding pair in the center, a nonbreeding adult male at each end, and other members of the cooperating group between them. Common Babblers and Yellow-eyed Babblers, also of India, likewise roost in contact in rows, up to twenty of the latter packed together, all facing the same way. In Australia, Orange-winged Sittellas sleep with a vigilant adult male at each end of the line, other members of the cooperatively breeding group toward the center.

Cooperative breeding is more frequent in the island continent than elsewhere, or at least has been more often recognized and studied. Many, if not all, species of the lovely little blue, or fairy, wrens nest in cooperating groups, some of which have been found roosting pressed together in rows. Splendid Blue Wrens sleep lightly and readily fly from their perch if disturbed, but with persistence John Warham succeeded in photographing eight of them on a slender twig of a low, bushy *Melaleuca* growing in an open paddock, all so closely packed, with heads hidden amid outfluffed feathers, that only their long, projecting tails could be accurately counted. In another family of birds that often have nest helpers, he found thirty Dusky Wood-

Long-tailed Tits on winter roost

Swallows sleeping side by side 40 feet (12 meters) up in a dying tree. Instead of lining up on a twig, other wood-swallows, such as the widespread White-breasted, cling together in a ball-like mass on a branch. If disturbed, the ball "explodes," much in the manner of a ring of sleeping Common Bobwhites, probably alarming a hungry nocturnal prowler. Warham also found about fifty Rainbow Bee-eaters, members of a family in which cooperative breeding is frequent, roosting in trees at a street corner of a small Australian town. In the evening they assembled with much squabbling, but after they settled down eight or more sat facing the same way with bodies touching, while a little apart a pair rested side by side. Another Australian bird that breeds in groups which sleep in contact is the black White-winged Chough, a mud-nest builder with a highly developed social life.

In northern Eurasia, Long-tailed Tits, who frequently feed young at nests not their own, sleep pressed together on branches, from nine to twelve in a row. In cold weather, they conserve heat by clumping into a compact ball, with tails sticking out in all directions. In the rain forests of northern South America and southern Central America, big, black Purple-throated Fruitcrows, who also breed cooperatively, sleep closely side by side on a high branch. Birds not known to be cooperative breeders that sleep in contact include Red-billed Toucans, of whom Geoffrey Bourne's flashlight picked out six roosting pressed together in a row on a high limb in the forest of Guyana.

In Africa south of Sahara, Speckled Mousebirds, also called colies, sleep clinging upright to the topmost branches of a small tree or bush, six or more in a compact bunch, their long tails hanging almost vertically beneath them. Sensitive to cold, they huddle together in the same manner on chilly and wet days. The tiny, short-tailed, largely green-and-red hanging parrots of the genus *Loriculus,* widespread from southern India to Indonesia and the Philippines, sleep in rather similar fashion. In tight clusters, they hang head downward, pendulumlike, in leafy trees. When they awake, they caress one another while remaining in this inverted posture. Their habit of sleeping upside down appears to be related to the way they frequently eat fruits or sip the nectar of flowers while clinging head downward. They share with lovebirds of the genus *Agapornis* the unusual habit of carrying amid the feathers of their rumps the materials that line their nest cavities. They are reported frequently to become inebriated by drinking too much fermenting coconut toddy from the vessels in which it is collected.

Birds who sleep in contact at night frequently rest in the same intimate manner by day, preening each other. They often feed their companions. If one is caught, they gather around, threatening the captor, or try to lure the captor away by distraction displays.

Communal Roosts

The roosting of some of the more gregarious land birds is a phenomenon so spectacular that it attracts the attention of people with no special interest in nature. Perhaps no kind of bird has ever slept in trees in greater multitudes, or more compactly, than the now extinct Passenger Pigeon of the deciduous forests of eastern North America. Their stupendous flocks, which for hours darkened the sky as they flew rapidly overhead, settled at nightfall to rest in the woodland, where they crowded upon the boughs in such incredible masses that, according to graphic contemporary accounts, thick limbs and whole treetops broke under their weight and bore many to their deaths on the ground. Such behavior in a highly successful species, evidently well adjusted to conditions prevailing in the primeval deciduous forests, is certainly most surprising. I surmise that it was a consequence of the rapid destruction of these forests by the axes of settlers of European origin advancing westward, causing flocks of pigeons that would have scattered more widely for roosting to concentrate disastrously in shrunken stands of trees.

Of extant pigeons, probably the Eared Dove in Argentina most closely approaches the extinct Passenger Pigeon in the size of its roosting aggregations. Millions of these colonial doves roost and nest in the same tangled, thorny, second-growth thickets, amid cultivated fields of the Province of Córdoba, covering the ground with their droppings. For a distance of about 37 miles (60 kilometers) they spread over the surrounding country to eat grains, especially sorghum, returning in the evening to their roosts. Unlike the pigeons that inhabit lower levels of tropical forests, such as the Gray-chested Dove, many that frequent treetops or open country roost communally, as will be told beyond.

Many parrots forage by day in chattering flocks and roost gregariously, although never, as far as I can learn, in such concentrations as a few kinds of pigeons do. In Australia, sometimes called "the land of parrots," John Warham watched colorful, red-breasted Galahs gather to roost. After their evening drink, they gradually drifted in small groups toward a patch of gum trees on a hillside. In the twilight, they indulged in wild aerobatics, singly or in pairs flying in and out and around the trees. Finally, they perched, often close together, about 40 or 50 feet (12–15 meters) up in the treetops, a few on dead branches, most hidden amid foliage. The few that could be seen perched with their heads tucked into the feathers of their shoulders. In strong contrast to the Galahs' vespertine play, the small flocks of Brown-hooded Parrots that I have watched go to rest in Costa Rica flew swiftly and directly into the crown of a tall tree in a clearing or at the forest's edge and, becoming silent, promptly vanished. At dawn they shot out and away with as little fanfare as when they entered.

In semiarid northwestern Costa Rica, as in arid Australia, parrots are more conspicuously abundant, and easier to watch, than in rainforested country. On the outskirts of a small town in Guanacaste Province in November, I found about twelve pairs of big, green Yellow-naped Parrots roosting in two low, spreading trees standing close together. The members of a pair slept side by side; two who might have been mated were about a foot apart; while one bird was quite alone. These parrots preferred to roost on the terminal twigs at the outside of the trees' crowns, where they were exposed to the sky and readily visible from below. At dawn, they flew over the town in pairs, a male and female closely side by side, as they had slept. On a remote hacienda in the foothills of the mountains in the same region, Crimson-fronted Parakeets flew in great, noisy flocks, and slept in the midst of heavy clusters of coconuts at the summits of some tall palms in front of the house.

Another parrot roost that I have seen was also in a dry region. One July evening, while loitering along the attractive waterside park of Guayaquil, Ecuador, enjoying the refreshing breezes that blew in from the broad Guayas estuary at the end of a warm day, the busy river traffic, and glimpses of distant Andean summits, a chorus of loud chirping drew my attention to a spreading *Erythrina* tree in front of the monument that commemorates the historic meeting of South America's two famous liberators, Bolívar and San Martín. Here, with much chattering and excitement, a flock of green Pacific Parrotlets, who had flown in from surrounding scrubby lands, was settling down to rest. And here they slept through the night, apparently undisturbed by nearby electric lights and the conversation of townsfolk passing beneath them on their evening promenade. When I returned to Guayaquil four months later, the little parrots still roosted in the same tree. Many another similar tree standing solitary in the surrounding countryside might have provided a more tranquil roost, but would they have been so free from nocturnal predators?

William Beebe described in detail how unidentified parakeets went to roost in tall bamboos in Guyana. First, they flew in small companies into the crown of an exceptionally high tree. After many had gathered there, the flock of several hundred birds, all screaming their loudest, rose like a whirlwind, ascended high in the air to trace several magnificent circles, half a mile to a mile in diameter, and spiraled down to a lofty tree near the bamboos. After a brief pause in this way station, the crowd of parakeets rose again and pitched into the bamboos, where they promptly fell silent for the night. Sometimes other large flocks followed the first. It is significant that the clump of bamboos was near the bank of a broad river and close to the house on a rubber plantation, where predators were probably less active than in the forest. Like the parakeets, many birds that sleep in large companies assemble at a neighboring point, often a tree, before they procede *en masse* to their roost.

The most populous roosts of flycatchers that I have seen in tropical America were of migratory rather than resident species. At Cañas, the small town in Guanacaste where Yellow-naped Parrots slept in pairs on the outskirts, scores of wintering Scissor-tailed Flycatchers roosted in its very heart, in tall orange trees behind the *jefatura,* or administrative headquarters. By day these exquisitely graceful birds were scattered over the surrounding country, showing little sociality. Toward sunset they were to be seen flying into the village from all sides, high above the roofs of the low buildings. Soon they began to settle in numbers in the tops of the orange trees, but, alarmed by somebody passing beneath them, or seized by sudden unrest, they would dart out and circle in the air, revealing on their sides the same tints that glowed in the roseate western sky. Before the glow had faded from the clouds, all were quietly settled amid the foliage, almost screened from view. With them roosted a few resident Tropical Kingbirds. In a later year, I watched Scissor-tailed Flycatchers streaming over the buildings of San José, Costa Rica, as the city lights went on in the evening. Undeterred by the noisy traffic and its fumes, they alighted in the trees of the central park to roost through the night. In roosting gregariously while in Central America, Scissor-tailed Flycatchers continue a habit begun in late summer in the southwestern United States, before their southward migration.

An orange tree with dark, sheltering foliage and thorns that may discourage predators, is a magnet that, as day ends, draws birds from a wide surrounding area. At Buenos Aires de Osa, a village in southern Costa Rica, many Fork-tailed Flycatchers slept in orange trees beside the little church. By day these sprightly birds, closely related to the Scissor-tails, spread afar over the neighboring savannas, where they perched low amid the grass, their white breasts resembling at a distance great snowy blossoms of some prairie herb. Each evening, as the sun sank low above the forested crests of the hills beyond the Río Grande, all the Fork-tailed Flycatchers in the vicinity, it appeared, came streaming in toward the village. The scores, possibly hundreds, of restless birds defied counting. If I appeared at the edge of the porch of the priest's house, where for a week in December I was the solitary occupant, or on the ground beneath the orange trees, while the birds were arranging themselves for the night, many would shoot out on all sides, their long, slender tail feathers whistling loudly as they rushed through the air. As soon as I made myself less conspicuous, they would return to their orange trees. After dark, I could move freely around them without disturbing their repose.

Swallows differ greatly in their ways of sleeping. Some permanently resident tropical species rest throughout the year in pairs or families in sheltered nooks; other swallows gather in vast flocks to roost. Few birds congregate in greater numbers; few indulge in such spectacular maneuvers as they approach the places where they will sleep. In North America all the common, widespread swallows—Barn, Cliff, Bank, Tree, and Violet-green—have been found in crowded roosts, either of one species or several together. The great evening gatherings of swallows are most spectacular in late summer and autumn, when young of the year are on the wing, and scattered families draw together as they prepare for their long migrations. They begin to assemble in July, sometimes during the first week, contain the greatest number of birds in August, and melt away in September, as the flocks move southward. "Thousands and thousands," "millions" are rough estimates of the size of a single gathering. This is often situated in marsh vegetation, among cattails, iris, water-oats, or reeds; but riverside willows, or a grove of trees in a small seaside town, may be chosen for the communal roost. Tree Swallows sometimes rest on the ground amid coastal dunes, as well as upon neighboring bushes, trees, fences, haycocks, and even roadways. Frequently, migrating swallows roost near water, a river, lake, or pond, where they drink or bathe before retiring. Year after year, swallows gather to sleep in the same place. Returning in spring, they again roost communally before they disperse to nest.

In the Sonora Desert of southern Arizona, where Purple Martins nested in holes that woodpeckers had carved in the giant Saguaro cacti, they roosted gregariously through the summer. In late afternoon, large numbers assembled on electric wires, before at dusk flying *en masse* to roost in cottonwood trees. In June and July, while female martins incubated their eggs and brooded their nestlings, males greatly outnumbered them. After the young began to fly, the size of the roost increased to about thirteen thousand birds in mid-September, after which the martins migrated southward until all had gone by early October. This roost was the focus of martins that by day were scattered over an area estimated by Milham Cater to be about 550 square miles (1,425 square kilometers).

In the Transvaal of southern Africa in March, an estimated one million migrating Barn Swallows, with very much smaller numbers of Cattle Egrets and other birds, roosted amid 10-foot (3-meter) reeds that densely covered a marsh about 600 yards (550 meters) in diameter. Their exodus next morning was explosive.

George K. Cherrie described the evening gathering of resident swallows that he witnessed along the Orinoco River:

> On the evening of the 19th of July, 1898, half way between Caicara and Altagracia, I had made my canoe fast in a tree-top, above one of the many submerged islands that are so common in the Orinoco, at that season of the year. A storm was gathering and it was near sundown, we were too far from either shore to attempt to reach solid ground for a camp. But the bird drama I witnessed that evening amply repaid me for the night spent in the tree-tops. Just before darkness I noted immense numbers of [Gray-breasted Martins, White-winged Swallows, Black-collared Swallows] and perhaps other species, arriving at or above one of the nearby islands of green tree-tops, where already there seemed to be tens of thousands of birds wheeling and circling about. The great masses of winged bits of life seemed to be influenced by a single mind, rolling like a wind-driven storm cloud, first to one end of the island and then to the other. Now rising high in the air, the next moment dropping almost to the tree-tops,

> then rising and circling again, the moving mass would resolve itself into a living cone descending rapidly point downward with a roar like a whirlwind. During this movement thousands appeared to drop into the tree-tops, then all orderly formation would be lost and the remaining multitudes returned to the rolling circling mass that marshalled its forces for another plunge toward the tree-tops. Darkness and the black, angry clouds of the coming storm hid the last acts of the bird drama and we crept beneath the *carosa* of our dugout canoe, where protected from the storm we were soon lulled to sleep by the rocking of the boat. (Cherrie 1916, pp. 161–162)

The inverted-cone or funnel-like formation that swallows assume, as they drop from their shifting, cloudlike aerial aggregations to their roost far below, has been recorded for a number of the more gregarious species. In April, I have watched huge flocks of swallows, migrating northward through southern Costa Rica, seek a place of rest at evenfall. Although by day they traveled always westward, following the direction of the coast and the principal mountain chains, wave succeeding wave in open formation, after sunset they ceased to hurry onward, and drew together into great compact masses that swung back and forth over the valley, now to one side, now to the other. Now and again this protean cloud would elongate downward into a spout or funnel, whose point would be directed to a certain group of trees, perhaps a clump standing upon a ridge, or a huge fig growing alone in a pasture; and many of the swallows would appear to dart down into the foliage. But then the cloud of circling birds, seeming nowise shrunken by these desertions, drifted far away. Because of the distance, the rapidly waning light, and the vast multitude of birds that I watched, I could never satisfy myself that any remained in the trees into which so many appeared to drop, nor could I discover where the great bulk of the migrants went to roost. Barn and Cliff swallows, who passed in immense numbers at this season, together with a few Bank Swallows, were probably all represented in these clouds of roost-seeking birds.

The valley where I write lacks marshes where swallows might roost, but our fields of sugarcane offer a good substitute for the reeds and other marsh plants of regions less thoroughly drained. In them sleep the Southern Rough-winged Swallows, our most abundant member of the family. After Blue-and-white Swallows have gone to rest in pairs or families in nooks beneath the roof tiles, the Rough-wings, lacking such sheltered bedrooms, continue to hawk for insects in the evening air, now circling high, now skimming low over pastures, each going its own way in loose formation. After the orange and lavender tints of sunset have faded from the clouds, they converge into a more compact flock, which swings back and forth over the valley and enclosing hills. The light is waning fast. Contracting into a denser mass and rising higher, the swallows form a dark cloud of constantly changing shape, in which each component particle continually shifts its position—a gigantic aerial amoeba with its internal granules in more than usual agitation, swinging hither and yon high above the darkening valley, dipping and rising once more. Finally, the huge amoeba begins to stream downward, each included particle shooting rapidly earthward, as though the ectoplasm had ruptured and, violently contracting, shot the contained particles toward the ground. Twice I watched this rapid downstreaming of the swallows, from elevations on opposite sides of the river, without being able to detect exactly where it ended, for the light was already so dim that the birds, easily distinguished as dark specks against the sky, vanished the moment their background became the dusky foliage of the trees. But at last I traced the downward movement to a small canefield, where I found a number of the swallows resting upon broad, nearly horizontal sugarcane leaves.

When I watched from a distance, the swallows appeared to shoot directly down to their destination, so swiftly that I wondered how they avoided dashing themselves against the ground. But one evening, hiding among the canes, I saw that, instead of streaming down directly above me, they flowed earthward some distance to one side, where I lost sight of them. Soon they reappeared, shooting into the tall canes all around me, flying a nearly horizontal course, and alighting easily upon the nodding cane leaves where they slept. Before sunrise in the morning, they sometimes rose up together from the canefield in a fairly compact flock. On other mornings, they departed their roost in more scattered formation, over a longer interval.

Swallows sometimes sleep in unexpected places. Near the end of the single long pier of the Pacific port of Puntarenas, Costa Rica, was a beacon light, mounted on a steel tower 40 or 50 feet (12–15 meters) high. On the evening of July 5, 1939, while the ship on which I was to sail lay unloading at the pier, I noticed several dozen swallows fluttering around the beacon in the rapidly fading daylight. Soon they settled on the rail that surrounded the light, already sending its fleeting red flashes across the broad expanse of the Gulf of Nicoya. At intervals a bird, quitting its place on the railing, would flutter against the glass cylinder that enclosed the light. The swallows chirruped much as they settled down; and from their voices, as well as their size, I identified them as Gray-breasted Martins; although the daylight was already too dim, and that of the beacon too dull and transient, to reveal their colors. Among the resident martins was a single Barn Swallow, unmistakable when its deeply forked tail was silhouetted against the sky. Most of its kind had already sped northward. After the last glow of daylight had faded, thirty or forty swallows rested motionless on the railing. A few tucked their heads among their feathers, but most kept their heads exposed; even after night was well advanced, I caught the gleam of their eyes in flashes of the beacon. Here they roosted with people walking and talking in several languages beneath them, while others fished from the end of the pier; with a steam locomotive noisily shunting about freight cars that served the unloading ships, sometimes enveloping them in clouds of sulphurous smoke. They could hardly have chosen a roost more inaccessible to predators, or noisier.

Among crows and their allies, communal roosting in trees is widespread and frequently spectacular. In autumn and winter, American crows gather from miles around to sleep, year after year, in the same sites, some of which are known to have been occupied for fifty years, or even a century. Two very large roosts, estimated to contain from 150,000 to 200,000 crows, were in past years located near Washington, D.C.; more often they are much smaller. The birds may gather amid dense conifers or other trees on the mainland, but often they choose a small island in a river, where they may rest only four or five yards up in a thicket of willows, or even lower amid dense reeds. Long ago, at such a roost, the Delaware River, swollen by heavy rain, rose suddenly in the night and drowned thousands of resting crows. When the weather became intensely cold and the wind blew fiercely from the north on an island in the Mississippi River opposite St. Louis, a company of crows, to avoid freezing their feet, abandoned a roost among willows to pass the night facing into the gale, in slight depressions on a snow-covered sandbank in front of the trees—somewhat as grouse and other birds seek warmth amid snow.

Black crows or their relatives, streaming in long, straggling flocks across barren fields to their roosting groves, as the sun sinks low on short winter days, are a familiar sight in northern lands. In spring and summer in the British Isles, adult Rooks of both sexes, as well as fledged young, sleep in their rookeries where they nest. In autumn they begin to congregate in larger numbers, those from many nesting colonies uniting in a winter roost, which is often the site of one of the rookeries. Occupied winter after winter, one of these communal roosts may contain thousands of Rooks, concentrated there from an area of perhaps 100 square miles (260 square kilometers). On most nights, dominant Rooks usually claim the highest positions in the trees, where they are safest from flightless predators and least likely to have their plumage fouled by droppings from birds above them; but when nights are extremely cold, they choose lower boughs, less exposed to wind. Often Jackdaws, who nest in holes rather than in arboreal colonies, share a roost with the larger Rooks.

Gregarious roosting is not restricted to crows of northern lands, for on an island near the coast of Ceylon (now Sri Lanka) House Crows gathered at night in palm trees, which by day were occupied by nocturnal fruit bats, who at dawn streamed in past the departing crows. Other corvids known to sleep communally, but in scores or hundreds rather than the thousands of roosting American Crows and Rooks, include Carrion Crows, Hooded Crows, Common Ravens, Pinyon Jays, and magpies. As they congregate in the late afternoon or evening from a wide surrounding territory, these birds do not always fly directly into their roosts. Frequently they collect in great numbers at some nearby station, in trees or sometimes on the ground, before they fly, often tumultuously, to the trees where they will pass the night, much as did the parakeets watched by Beebe in Guyana.

Among thrushes, American Robins often roost in large aggregations, sometimes in swampy woods where broad-leaved trees form a thick canopy overhead, in shade trees in the center of a town, or in a thicket of ancient lilac bushes, as in William Brewster's garden, close by his house. These roosts are at first occupied, as early as May, by adult males, apparently including paired birds, whose mates still incubate or brood nestlings. After the young fledge, the male robins are joined by females and juveniles, sometimes swelling the roosts to an estimated twenty-five thousand robins. Other birds who often sleep with the robins are Common Grackles, Brown-headed Cowbirds, Red-winged Blackbirds, Eastern Kingbirds, Baltimore Orioles, Cedar Waxwings, and Brown Thrashers. Redwings (thrushes) wintering in northern England gather nightly from a wide area to roost together.

From Ireland to Japan, Pied, or White, Wagtails sleep communally. Early in the season, the birds at these roosts are largely adult males, whose mates may be nesting. Soon they are joined by females and juveniles. The most famous of the wagtail roosts, in plane trees in the center of O'Connell Street in the city of Dublin, at one time sheltered about 3,600 birds. In the evening they began to arrive silently, but as the crowd swelled they burst into a twittering chorus, audible for nearly 100 yards above the noise of traffic.

Among the best-known of all roosts are those of European Starlings, who often in their thousands become a scarcely tolerable nuisance by fouling urban buildings and sidewalks. In autumn, when the local breeding population is augmented by unattached juveniles, migrants from other parts of Britain, and immigrants from the continent of Europe, starling roosts in England may be crowded with a hundred thousand to half a million birds, each of whom tends to rest on the same perch night after night. As they go to roost in the evening, these thousands of starlings often sing together, or with a great roar of wings they rise all together to perform spectacular, highly coordinated maneuvers high in the air. V. C. Wynne-Edwards learned that the members of each roost foraged over a particular area that he called a communal territory, not sharply delimited but overlapping adjoining territories. Starling roosts in the British Isles are known to have been occupied, with brief intermissions, for well over a century. While the majority of the starlings flock to the communal roosts, sometimes from foraging grounds as far as 30 miles (50 kilometers) away, mated pairs with dependable sources of food remain aloof from these assemblies and sleep through the winter in holes and nest boxes where they have nested. In the darkness of a winter morning, they sing softly before they emerge. In late afternoon, they sing again while sitting conspicuously on chimney pots, watching for flocks to fly past on their way to the crowded roosts.

A study by Joseph F. Jumber in central Pennsylvania revealed how European Starlings change their roosts with the changing seasons. In mild summer weather, a multitude slept mainly in heavily foliaged deciduous trees, especially Norway Maples and Sycamores with dense canopies, scattered through the city of State College. At the approach of autumn, when some of these starlings would migrate from these dispersed roosts, they collected nightly in a single grove of about twelve deciduous trees, all different from those occupied in summer. By October's end, the starlings who did not migrate abandoned this autumn roost to sleep in more sheltered situations, including ledges of city buildings, warehouses, airplane hangars, bridges, and electrical substations. When these preferred sites were lacking or overcrowded, the starlings might roost on bare deciduous trees well protected by buildings or sheltered in groves of evergreens. In the surrounding country, many slept in barns. In early spring, when migrants began to join the resident birds, the starlings of State College again gathered in a single large roost, this time in a dense grove of evergreen trees about 6 miles (10 kilometers) from the city, where they mingled with Red-winged Blackbirds, Common Grackles, Brown-headed Cowbirds, American Robins, and Mourning Doves. After nesting started in late May, the spring roost disintegrated, and about five thousand of the starlings moved to an early summer roost amid cattails intermixed with low shrubs in a marshy area. By mid-June they had abandoned this location to return to the summer roosts in the shade trees of the town. These starlings exhibited surprising versatility by shifting to very different sites to meet the stresses and opportunities of the changing seasons.

The evening gatherings of these starlings proceeded in four stages. About an hour before sunset, they began to collect in small flocks on their foraging grounds and started to move toward the roost, frequently stopping to eat along the way. As the Sun dropped lower and they flew toward the roost along definite, established routes, they were joined by other incoming starlings, swelling the homing flocks to thousands. Instead of flying directly into the roost, the returning birds settled in a preroosting assembly on nearby trees, buildings, power lines, or other conveniently situated perches, from which, as daylight waned, they passed in growing numbers to the trees or buildings where they would sleep.

Many species of the troupial family, often called icterids, not only roost gregariously but likewise forage socially and nest in colonies. An estimated fifteen million Red-winged Blackbirds gathered in a tremendous roost in the Dismal Swamp of Virginia. Like Rooks, dominant Red-winged Blackbirds claim the best positions for themselves. At a small roost in a cattail marsh, the older males rested toward the center of the dense vegetation, where they were more sheltered and safest from predators, leaving the juvenile males to sleep amid sparser vegetation at the edge. These blackbirds often roost with Brown-headed Cowbirds and European Starlings. Chattering flocks of Common Grackles winging roostward in the evening are a familiar sight in eastern North America. Similarly, Great-tailed Grackles, who by day spread widely over cultivated fields and riversides in northern Central America, converge in the late afternoon upon the central plaza of many a small town and village, to sleep in palms and other trees.

Years ago, I passed a season on a banana plantation on the border between Guatemala and Honduras, where Great-tailed Grackles roosted in the coconut palms that surrounded the overseer's house. A lively scene they presented, as at sunset they flew up from the river, along whose shores they foraged in the late afternoon, to the hilltop where the house stood. After much shifting around, they finally settled down for the night among the inner fronds of the palms. From February to June, the female grackles nested in the crowns of the same palms where the flock slept. This was a bad arrangement. The males, who in this polygynous or promiscuous species do not attend nests, and the females not at the moment engaged in incubation or brooding nestlings, together always outnumbered mothers busy with eggs or young. When the flock returned to roost in the evening, the confusion and disorder in the palm trees was so great that I wondered how incubating females managed to stay on their eggs. Angry cries, which at this time arose from birds unseen amid the palm fronds, were doubtless the complaints of protesting hens. I attributed to the grackles' disorderly habits the loss of many of their eggs and nestlings, for they drove away every other large bird that approached their colony. Moreover, droppings of the roosting birds fouled the fronds around the nests.

Other colonial icterids, such as oropendolas and caciques, whose long woven pouches cluster in isolated treetops, arrange things better. All males, who never help at the nests, and all females not incubating or brooding, invisible in their pouches, go elsewhere to sleep, so that after nightfall the colony appears deserted. The roost of a flock of Montezuma Oropendolas in the Lancetilla Valley of northern Honduras was at first an avenue of tall bamboos, about half a mile from the tree where the nests hung. Here they slept with Chestnut-headed Oropendolas and a number of Garden Thrushes, also called Clay-colored Robins. Apparently because I stood too often and conspicuously to watch them retire in the evenings, the shy Montezuma Oropendolas moved to a hillside about 2 miles (3 kilometers) from their nesting colony. Thither they flew in successive flocks late in the evening; thence they returned to their nest tree in the gray light of dawn, flying up the narrow valley with wing-strokes as regular as crows', and appearing as large and black against the sky.

Even in their breeding season, Melodious Blackbirds, whose nests were scattered singly, gathered in numbers to roost in a dense stand of Giant Canes beside the Río Morjá, a tributary of the Motagua in northeastern Guatemala. Here they were joined by their relatives, Bronzed Cowbirds, almost as black as they, except their red eyes. Nearby among the tall canes slept wintering Orchard Orioles. I knew no more delightful way to end the day than by sitting on the sandy shore beside the canes, listening to the clear, soothing whistles that issued from among them. As the Melodious Blackbirds grew drowsy, their notes became lower and more widely spaced, until in the deepening dusk they fell asleep and were heard no more.

In arid regions of Africa south of the Sahara, Red-billed Queleas that forage over the savannas gather by hundreds of thousands and millions to roost in thorny trees and shrubs. When a vast horde of Bramblings migrated from boreal forests to central Europe in the winter of 1950–1951, the almost incredible number of seventy-two million were estimated to gather in only two roosts. In tropical America, another small granivorous bird, the migratory Dickcissel, which winters chiefly where rice is grown, roosts in aggregations that are impressively large, although nowise approaching those of the African weavers or the Bramblings. On the island of Trinidad, Richard ffrench found Dickcissels sleeping in bamboos and fields of sugarcane, which if distant from their foraging grounds they often approached flying in one continuous column twenty or thirty birds wide and up to a mile long, resembling from a distance a swarm of locusts. Inexplicably, to reach the preferred roosting site they sometimes passed for up to 15 miles (24 kilometers) over almost continuous stands of cane that appeared equally suitable for their nightly repose. Amid the cane, which grew to be 10 or 12 feet (3–4 meters) high, up to a dozen birds rested in compact rows on horizontal lengths of the curving leaves, at least 6 feet (2 meters) above the ground. After settling on the cane leaves, the Dickcissels called all together, producing a highly amplified hissing sound that on still evenings was clearly audible three-quarters of a mile (1.2 kilometers) away.

Inevitably, such large aggregations of resting Dickcissels, sometimes approaching 100,000 birds, attracted predators. Among those seen at the roosts were the mongoose and domestic cat, Aplomado Falcons, Barn Owls, and the Merlin or Pigeon Hawk. Although attacks by Merlins were often unsuccessful, they appeared to subsist largely upon Dickcissels. A widespread practice of sugarcane planters is to burn off dead leaves and other trash immediately before cutting the canes. Usually the smoke, the flames, and the crackling sounds they made aroused the sleeping birds in time to save themselves; but if fires were lighted simultaneously around the whole perimeter of a field, the birds might be trapped and perish miserably.

As far as I have learned, no other emberizid or fringillid finch of the New World roosts in such vast multitudes as the Dickcissel. About 200 Purple Finches roosted in winter in densely foliaged cedar trees in Tennessee. Some forty Evening Grosbeaks rested in a grove of tall pines in Ontario. Twenty-four Lesser Goldfinches settled for the night in a leafy cottonwood tree in Arizona.

Roosts of Mixed Species

Among the land birds of the American tropics, especially its forested regions, huge aggregations of sleeping birds are rare. Here, where many of the resident birds live throughout the year in pairs on or near their nesting territories, and many provide shelters for themselves, flocking is not so widespread and conspicuous as among birds of regions where a severe winter causes the disintegration of families and forces many birds to seek food far from where they nest. The foregoing accounts contain several examples of mixed roosts, as of Brown-headed Cowbirds and European Starlings with Red-winged Blackbirds, and several species with American Robins. In the Colusa Marsh of California in autumn, at least several million Red-winged Blackbirds, Tricolored Blackbirds, Brewer's Blackbirds, Brown-headed Cowbirds, and European Starlings slept together.

The roosts that I have seen in the tropics were more remarkable for the variety than for the number of birds that resorted to them. One of the most interesting was in a small patch of introduced Elephant Grass, more than head high, in a cleared valley in northern Honduras. This was the nightly sleeping place of most of the small seedeaters of the vicinity. The little, black male White-collared Seedeaters, among the smallest of finches, easily outnumbered all the others. With them were three other dusky species, the all-black race of the Variable Seedeater, the equally black Thick-billed Seed-Finch, hardly to be distinguished from the former except by its thicker bill, and the shining Blue-black Grassquit. The brownish or grayish females of these four small finches roosted here, too, but as long as they were occupied with eggs or nestlings, they came in much smaller numbers than their consorts. An hour before dusk these birds would begin to congregate, arriving singly or in pairs or in groups of four or five, flying low over the open field until they swerved upward into the tall, coarse grass. Here they perched, keeping up a continuous chatter of small voices until well after sunset; and now and again a burst of canarylike song, arising at one end of the long patch of grass, would be joined by a hundred throats and sweep like a gust of wind from end to end of the assemblage, then die as suddenly away.

With the seedeaters slept a family of Black-cowled Orioles, whose nest had been suspended beneath the giant leaf of a Traveler's Palm not far away. They were joined by migratory Orchard Orioles in August, and in October by Baltimore Orioles, who were never numerous. Before the Baltimore Orioles came the Eastern Kingbirds, midway of their long migration from North America to the southern continent. Here they interrupted their journey to linger for over a month, as they sometimes do on both their southward and northward flights. Surprisingly, they chose to roost in the tall grass with the seedeaters and orioles; one would have expected these birds of high, open spaces to prefer loftier quarters for the night. Last of the motley assemblage to disappear into the high grass, they traced intricate courses as they snatched insects from the evening air before they dropped quietly into it, the only silent figures in all that loquacious throng. I never noticed any bickering among the eight species of diverse habits that crowded into this roost.

Of the hundreds of species of plants introduced into the New World tropics, none is more attractive to roosting birds than orange and other citrus trees and timber bamboos. A tall, dense clump of these bamboos, close beside a house on a great coffee plantation in Guatemala, offered lodging for as cosmopolitan a group of travelers as ever gathered in a wayside inn. As the sun set on January evenings, a dozen or more wintering Rose-breasted Grosbeaks, the males now with only a rosy tinge on their light breasts, darted in among the close-set, leafy stems so swiftly that I could not count them accurately. Several Baltimore Orioles of both sexes plunged into the clump with equal abruptness, and with them a few Orchard Orioles. Less conspicuous were the little olive-and-gray Tennessee Warblers that went to roost with them. In addition to these vacationists from northern lands, several local residents put up at this inn, among them Yellow-winged Tanagers with lavender-tinged heads and Garden Thrushes, or Clay-colored Robins, dressed modestly in brown. Unlike all the other guests, who sought their lodging as unobtrusively as possible, three Vermilion-crowned, or Social, Flycatchers would every evening, before retiring, perch in full view on the tops of the bamboo shoots and repeat their plaintive notes many times over. A female Great-tailed Grackle, feeding nestlings so early in the year in the same bamboo where many birds slept, continued to bring food to them in the waning light, after the others had settled down to rest.

Another notable roost was in the hamlet of Buenos Aires de Osa in southern Costa Rica, where Fork-tailed Flycatchers slept in orange trees. Beside the path that led up to the small, infrequently used church was a double row of well-grown *Dracaena fragrans,* an arborescent member of the lily family with stiff stems crowded with broad, close-set, parallel-veined leaves. Here in December slept at least a score of Baltimore Orioles, who around sunset arrived singly or in small flocks and promptly settled down among the bases of the leaves, which screened them so well that even the glowing orange of the adult males was invisible amid the dark green foliage. With them roosted a single Orchard Oriole, a rare winter visitant in this region, and at least one gemlike Golden-masked Tanager, whose mate probably slipped in unseen. Here also slept many Ruddy Ground-Doves, who by day spread over the surrounding fields. They were as difficult to count as the orioles but probably did not exceed forty. Nevertheless, this was the largest gathering of roosting pigeons that I have met in the humid tropics, where nothing approaching the great concentrations reported from temperate regions of North and South America has come to my notice.

In India as in America, mixed roosts are frequent. House Crows, Common Mynas, and Rose-ringed Parakeets sometimes sleep together. Rosy Pastors roost with Common Mynas, as do House Crows and Jungle Crows. At other roosts, mynas associate with birds as different as Cattle Egrets and Little Egrets. Mynas, House Crows, Jungle Crows, Rose-ringed Parakeets, and Cattle Egrets slept in a single Peepal tree in a well-wooded area.

Why Birds Roost Communally

What advantages do birds derive from roosting in large companies instead of alone or with their mates and, possibly, independent young? When they curtail their foraging time and spend energy flying many miles to a large communal roost, sometimes passing over unoccupied sites quite similar to their destination, these advantages must be substantial. What can they be?

The Pacific Parrotlets that roosted in a waterside park in Guayaquil, the Scissor-tailed Flycatchers who at nightfall flew into the central park of Costa Rica's capital, the Gray-breasted Martins who perched on a beacon tower on a bustling pier at Puntarenas, the Pied Wagtails that congregated for the night along one of Dublin's main streets hardly gathered in these situations to enjoy quiet repose. But in all these places they were less exposed to predators than they would have been in the surrounding open country where they foraged by day, and they were probably little molested by people, except by the noises they made and the fumes from their cars or engines. These birds traded tranquillity for safety.

Other roosts that I have seen were in relatively safe places, often in, or at the edge of, a town or village, or in isolated trees or groves less likely to be visited by climbing quadrupeds or snakes than are trees in closed woodlands. Orange trees, frequently chosen for roosting, have dense concealing foliage and forbidding thorns. Timber bamboos with hard, smooth trunks appear to be difficult for most quadrupeds to climb, although not for monkeys with grasping hands, and their twigs would yield under the weight of an approaching animal, waking the sleeping birds. Similarly, grasses and other herbaceous plants would sway if even a small snake tried to climb them.

One wonders how so many birds of different species, which do not flock together, find these safe roosts that are sometimes far from where they forage and nest. Probably after one or two exploring individuals have discovered them, others follow leaders who fly toward them in a confident manner. Derek Goodwin pointed out that the tendency to roost with companions is strongest in birds that are sexually immature, unpaired, or wintering at a distance from their breeding grounds; weakest in paired, territory-holding adults—a generalization amply supported by my experience in tropical America. Goodwin's studies of bird behavior suggested that many diurnal birds find approaching darkness slightly frightening, especially in a hostile or unfamiliar environment. In such a psychic state, a bird would more readily follow other individuals, of its own or some other compatible species, who appeared to know a safe place for sleeping. By such means, large roosting aggregations might arise.

Many birds gather to roost in a site that is intrinsically safe; whether it becomes safer as it becomes crowded is less obvious. Anyone who studies how birds pass the night knows how difficult it is to find those who sleep alone, how relatively easy to locate communal roosts. Birds flying toward a central point from various directions, often engaging in conspicuous aerial displays before they enter the roost, frequently squabbling noisily and shifting about restlessly as they settle down, must certainly reveal the location of their roost to every alert predator within hundreds of yards, if not within miles. When a site is occupied night after night and year after year for many decades, it may well become a traditional source of food for the predators of a wide surrounding area. What can compensate for this unfortunate publicity?

A first thought is that a site may be so much safer than any other for a long distance around that it attracts many birds, and, despite the fact that as their numbers build up predation increases, an individual's chances of surviving are greater than if it slept alone in a less secure situation. This may be true if the site is unique, such as the palm-shaded central plaza of a small town where Great-tailed Grackles assemble from surrounding open country, but it can hardly apply to Dickcissels in Trinidad, who fly over wide expanses of sugarcane, all looking much alike, to gather in enormous numbers in a certain canefield. Counterbalancing the conspicuousness of a great communal roost is the probability that at any moment some of its members, wakeful and alert, will detect an approaching enemy and cry out, warning sleeping neighbors of this peril; much as, in mixed flocks of forest birds, many watchful eyes and many voices quick to sound the alarm more than compensate for the conspicuousness of the party. A similar advantage might accrue from sleeping in mixed roosts. Although flocks of different species arriving at different times and performing different preroosting displays may make such roosts even more prominent than large roosts of a single species, diverse ways of sleeping may increase the probability of detecting an approaching enemy.

In an experiment reported by Amotz Zahavi, a bird-banding group at Reading, England, played the role of a ground predator trying to catch a Pied Wagtail sleeping in a large, compact group in a lighted area. Despite the illumination, they found it easy to approach a bird sleeping as little as a yard away from its massed neighbors. It was much more difficult to catch a wagtail from a compact flock, for one or another was often awake and when approached would fly away, with cries of alarm that started a chain reaction, stirring up the whole group. However, despite crowding, Dickcissels sleeping in the canefields of Trinidad suffered predation from Merlins, Barn Owls, and apparently several other predators that were seen there. Nevertheless, the victims probably represented only a small percentage of the many thousands of roosting Dickcissels. We do not know whether the total mortality would have been more or less if the same number of birds had been thinly dispersed through the great expanses of sugarcane on the island. In contrast to the situation in Trinidad, at a large communal roost of European Starlings in New Jersey, no evidence of predation was noticed.

Peter Ward and Amotz Zahavi believed that communal roosting evolved primarily for the efficient exploitation of unevenly distributed sources of food by serving as "information centers" for birds that forage in flocks. Different parties of birds such as Red-billed Queleas, European Starlings, Cattle Egrets, or parrots, who by day spread widely over the surrounding territory, may be diversely successful in finding productive areas. By gathering at nightfall into a common roost, these parties may pool information. At daybreak, when the birds disperse to forage, those that enjoyed abundance on the preceding day, without exhausting the supply, will fly confidently to the same place; those that fared badly may follow them. The preroosting displays, often high in the air and visible from afar, serve to advertise the location of the roost and draw other birds to it; for the greater the number of participants and the wider the area from which they come, the greater the probability that some of them will have found good foraging.

This putative function of the communal roost is hard to prove; the observation that at the morning exodus from a large roost some parties fly off promptly, as though they knew just where to go, while other birds delay in the roost or nearby, as if waiting for guidance, is the strongest, although inconclusive, evidence in support of the hypothesis. Ward and Zahavi regard all the measures taken for the security of the nocturnal gathering, such as the choice of a safe place, a system of warning, and joint defensive tactics, as adjuncts to the roost's main function of diffusing through a population information about temporary sources of food. In this view, the roost is essentially a cooperative assemblage, beneficial to the species—an idea repugnant to many evolutionists who strenuously reject group selection.

To test the "information-center" hypothesis, T. H. Fleming spread, on successive days, a gallon of maggots over the ground where individually recognizable Pied Wagtails repeatedly foraged. In none of five repetitions of the experiment were the birds who frequented the locality followed to it by newcomers who might have benefitted by this bounty. Although these trials lend no support to the hypothesis, they would need to be repeated many times with different species in diverse situations in order to disprove it.

By tracking European Starlings equipped with miniature radio transmitters in New Jersey, Douglas Morrison and Donald Caccamise discovered that these birds were more strongly attached to certain favored areas for foraging than to their roosting sites, which they changed rather frequently. The communal roosts were largest, with many thousands of starlings, in late summer in dry years when food was scarcest, which suggests that they might then serve as information centers.

The information-center hypothesis can be tested at nesting colonies of birds as well as at communal roosts. Both David N. Nettleship and Anthony J. Gaston found support for it at colonies of hundreds of thousands of murres in the far north, where rich concentrations of food are spottily distributed over vast expanses of seas partly covered by floating ice pans. Outgoing birds uncertain where to go alight on the water at the foot of their nesting cliffs, waiting for better-informed birds to lead them to more profitable foraging areas, or for flocks of incoming birds, who presumably have foraged well, to indicate by the line of their flight the direction in which abundant food might be found. Gaston believed that in the Canadian Arctic, colonies of less than ten thousand Thick-billed Murres could not persist for lack of enough birds to help each other locate food in constantly shifting expanses of open water amid ice-covered seas. Since sea-bird colonies grow by gradual accretions to a small nucleus of founders, the present huge colonies in the far north must have started at a period when the climate was milder and open water more extensive.

Support for the information-center hypothesis comes also from colonies of nesting Cliff Swallows, where Charles R. Brown noticed that a bird who has foraged unsuccessfully returns to the colony, finds a successful forager, then follows that individual, presumably to a concentration of flying insects.

A different explanation of the communal roost was advanced by V. C. Wynne-Edwards, who regarded safety, and even sleeping together, as subsidiary to the roost's primary function, that of drawing many birds together at a definite time, readily known without a clock, the evening when daylight wanes rapidly, so that they might engage in mass displays. By participating in these demonstrations, the birds could estimate their numbers in relation to the experienced resources of their habitat, and when these were inadequate, reduce the local population by emigration. Moreover, these assemblies might serve to move the migrants through a region in an orderly progression. This hypothesis derives some support from the observation that, at least in the American tropics, many migrants gather in large communal roosts.

These diverse explanations of communal roosts are not necessarily mutually exclusive, and none accounts for all the facts. For many roosts, safety of the site is evidently the determining factor. Although the information-center hypothesis is attractive, it can hardly apply to birds like Scissor-tailed Flycatchers who, after sleeping together, disperse over the countryside to forage. Even less can it explain why species of quite different diets gather at the same mixed roost. We should not overlook the possibility that vast aggregations of sleeping birds do not confer benefits proportionate to their size but result simply from the contagiousness of mass movements: spreading from a center, such a movement may attract more and more distant birds, until it is joined by birds who have foraged miles away, swelling the roost far beyond the size where the participants derive any advantage from its great numbers. The preroosting demonstrations, the gyrations of swallows, swifts, starlings, and other birds high above the roost site, may be, above all, expressions of the excitement generated by the meeting of great numbers of birds, or a display of exuberance. Or the birds may delay in the air, whirling about, because they hesitate to dive into a roost where peril may lurk beneath concealing foliage of an umbrageous tree or a canefield, until one or more bolder or more impetuous individuals start to plunge into it, thereby touching off a mass descent; much as Antarctic penguins line up at the edge of an ice shelf, fearing to jump into the water where voracious leopard seals may await them, until a more courageous spirit leads the way. Possibly birds gather in populous roosts because, like many people, they feel safer among a crowd. As long as they are in no greater peril than they would be in solitude, natural selection could not veto the satisfaction of this feeling. In any case, communal roosting, above all when its participants include birds who have not been closely associated in flocks throughout the day, is a complex phenomenon which baffles our understanding because the feelings of birds are hidden from us.

Perils of Communal Roosting

Becoming conspicuous to predators is not the only liability to which birds expose themselves by sleeping in large aggregations. In crowded roosts, birds are not always careful to place themselves where droppings from others above them will not fall upon them. Yoram Yom-Tov demonstrated experimentally that fouling by uric acid and feces decreases the water-repellent quality of plumage, making it much more absorptive of rain and less able to retain the birds' vital heat. For three weeks he kept European Starlings in a cage below another cage with starlings whose droppings fell upon them. Then for fifteen minutes he exposed these soiled birds to heavy artificial rain. Within half an hour, a third of the drenched starlings, each of whom had absorbed from 10 to 16 grams of water, succumbed to the wetting. The clean birds in the upper cage, deluged in the same manner, retained very much less water and survived the experiment. Among Rooks, Choughs, and probably other birds, older or dominant individuals who occupy the higher perches in communal roosts avoid this hazard to which their subordinates, resting lower, are exposed.

Birds that sleep in crowded roosts should be careful not to arouse human hostility, as by attacking crops, fouling buildings and their surroundings, killing trees with accumulated excrements, disturbing human repose with noise, or otherwise making themselves unwelcome, for with their explosives, poisons, and shotguns humans can readily destroy immense numbers of crowded sleepers. Crows, starlings, blackbirds, queleas, and others have paid dearly for their propensity to gather in great roosts. If this persecution, becoming widespread, is prolonged over many generations of birds, natural selection should favor individuals who roost apart or in small, inconspicuous parties, until the species most closely associated with humankind cease to congregate in great numbers. Thereby future human generations, spared from certain annoyances, would be deprived of some of the most impressive phenomena of the feathered world.

References: Beebe, Hartley, and Howes 1917; Bent 1946; Bent et al. 1968; Bourne 1975; Brewster 1949; Brown 1986; Bucher 1970; Caccamise, Lyon, and Fischl 1983; Cater 1944; Cherrie 1916; Emlen 1938; Erickson 1938; ffrench 1967; Fitch 1947; Fleming 1981; Gadgil 1972; Gaston 1977; 1978a; 1978b; 1987; Goodwin 1976; Greig-

Smith 1979; Hudson 1920; Johnston 1960; Jumber 1956; Laskey 1958; Meinertzhagen 1956; Morrison and Caccamise 1985; Nettleship and Birkhead 1985; Noske 1980; Orians 1961; Rudebeck 1955; Skead and Ranger 1958; Skutch 1954; 1972; 1981; Still, Monaghan, and Bignall 1987; Tennessee Ornithological Society 1943; Van Someren 1956; Ward and Zahavi 1973; Warham 1957; Weatherhead and Hoysak 1984; Wynne-Edwards 1962; Yom-Tov 1979; Young 1971; Zahavi 1971a; 1971b.

FIVE

Sleeping in Dormitories: A General Survey

Of the approximately nine thousand species of birds, relatively few sleep with better shelter than the foliage of trees and bushes. When we remember that the sleeping habits of birds have received inadequate attention, and we do not know how most birds pass the night, this might appear a rash, premature generalization. And so it would be, if it were not made in the light of certain guiding principles.

Let us designate by the term "dormitory" any sheltered place, other than a perch amid vegetation, the ground, or water, used by a fledged bird or birds for sleeping when neither incubating eggs nor brooding young. Dormitories include covered nests built by the birds themselves, holes in trees, whether carved by them or otherwise made, burrows in the ground, crannies in cliffs, nooks in buildings, and similar shelters. With this inclusiveness, we might say that all birds who sleep above the ground or water (except the very few believed to sleep in the air) occupy either a roost or a dormitory. Roosting birds, except those on open ledges, firmly clutch the perch with their feet; those that sleep in dormitories, which nearly always have firm floors, may lie with their ventral surface in contact with them. Certain birds have greater need of protection from below than from above. Accordingly, we include among dormitories the open cups or platforms built by Magpie-Geese, rails, coots, gallinules, and perhaps other marsh birds for sleeping and brooding their mobile chicks above water or wet soil, and likewise the bulky open nests made by Curve-billed Thrashers of the arid southwestern United States to shield their sleeping bodies from the sharp spines of the cholla cacti which protect them from nocturnal prowlers. With these few exceptions, dormitory nests are usually roofed or covered.

I know of no bird that habitually sleeps in a covered dormitory

Hairy Woodpecker clinging in front of dormitory hole

but hatches its eggs and broods its young in an open, cup-shaped nest placed amid foliage or in some more exposed situation. However, it does not follow from this that all birds who lay their eggs in roofed nests use such structures as dormitories. Many kinds of American flycatchers build snug, enclosed nests of the most diverse forms, but some of these nests that would make admirable dormitories are not used as such. Similarly, a number of finches employ their roofed nests only for rearing their young. Among the exceedingly diverse nests of cotingas and tityras of tropical America are bulky closed structures with a narrow orifice in the side and cavities in trees, yet I have found none of these birds sleeping in dormitories. The absence of nests of these birds in good repair, in the months when breeding is in abeyance, strengthens my belief that they are not used for sleeping. Accordingly, from the facts that only a minority of birds breed in covered nests of some sort, and that not all birds that lay their eggs in covered nests sleep in such nests, we may with confidence conclude that a smaller minority occupy dormitories. Most birds roost with no better protection from wind, rain, and nocturnal chill than foliage or a windbreak provides. However, a large tropical leaf may adequately shelter a small bird from hard rain; and at high latitudes some birds escape extreme cold by burrowing under snow.

It might be thought strange that a bird, able to build a snug covered nest, or to carve a cozy hole in a dead tree or clay bank, should not provide such a shelter to protect itself from cold or rain by night—or even by day. Few parts of the Earth, even in the tropics, are without many nights in the course of a year too cold or wet for comfortable sleeping out-of-doors, at least by creatures as sensitive to these discomforts as we are. But birds, like turtles, carry their houses wherever they go; they fly under their own roofs. Their individual dwellings are less cumbersome than the turtle's rigid, clumsy carapace; they permit greater freedom of movement, and they never hold a bird a helpless prisoner if, as rarely happens, it falls upon its back. Moreover, by molting, a bird might be said to build itself a new dwelling every year. The bird's feather-house has a water-repellent roof and walls—see how raindrops gather into shining pearls and roll off the plumage of a bird's back, anointed with oil from its preen-gland! By fluffing out its feathers so that they enclose more tiny air-spaces, the bird in effect surrounds itself with thicker insulating walls, or puts on a thicker coat, when it feels cold. By sleeking down its plumage, it increases heat loss when it is warm.

Moreover, against the advantages of sleeping in dormitories must be weighed certain disadvantages. They may be infested with lice or other vermin. A nest that is a comfortable abode in the dry season when eggs are laid may be made of materials that soak up and retain too much water when it rains. If attacked by a nocturnal marauder, the sleeping bird may not be able to escape swiftly enough through its dormitory's single doorway. Thus, the relatively huge constructions of castlebuilders, firewood-gatherers, and related birds have the disadvantage which, as the history of warfare clearly demonstrates, is inseparable from all strongholds, however massively built: the occupants may be bottled up in them by the enemy. Probably this is the reason why castlebuilders of the genus *Synallaxis* do not, as far as known, use their strongly built nests as dormitories (although certain related birds with equally well enclosed nests of interlaced sticks lodge in them), and Rufous Horneros roost amid foliage instead of sleeping in their thick-walled nests of hardened clay that resemble Latin-American baking ovens.

Natural cavities or holes carved in trees by other birds may not be numerous enough to provide dormitories for all individuals of a certain hole-nesting species; sometimes too few are available for a smaller number of breeding pairs. The decaying tree or limb in which a bird sleeps may fall in the night. The burrow in a riverside bank that safely shelters eggs and young in the dry season when the current shrinks may be flooded or washed out when the river rises. In view of these hazards of sleeping in dormitories, it is not surprising that birds, who live in feather-houses, do not more often laboriously construct additional dwellings for themselves.

One might suppose that the more severe the climate, the more widespread the construction of dormitories would become; that they would be more numerous among birds that winter amid snow and ice at high latitudes than among residents of warm tropical lowlands. The reverse is true. For rather obvious reasons, the dormitory habit is more widespread among tropical birds than among those that breed nearer the poles. In the first place, birds that perform long migrations, or in the inclement season wander in flocks in search of food, would gain little by constructing stationary dwellings for themselves. In Britain, resident European Starlings with sound nest cavities sleep in them in winter, while immigrants from the continent join unestablished local individuals in great communal roosts. In the tropics, a much larger proportion of the birds live throughout the year, in pairs or singly, on or near their breeding territories. At all seasons, these permanent residents have no lack of foliage to screen their dormitories, or they can build them in trees or shrubs infested by stinging ants that attack flightless invaders.

Dormitories would be too conspicuous in leafless deciduous trees in the northern winter. Evergreen conifers offer better concealment for dormitory nests, but at the same time make them less necessary because the dense foliage of some of them offers many birds welcome shelter from winter's gales. Even in the tropics, where many migrants from the north reside for half or more of the year, frequently on defended territories, these travelers never, as far as I know, build dormitories for themselves. I suspect, however, that the Yellow-bellied Sapsucker, the only northern woodpecker to winter as far south as Costa Rica, follows the family tradition of sleeping in a hole wherever it can find one. As day ended, on two occasions, a migrating sapsucker inserted itself, as far as it would go, into a depression in the side of a trunk too shallow to contain the whole bird.

One might also expect that more birds would sleep in dormitories on cool tropical mountains than in the warm lowlands. I looked into this matter during a year spent above 8,000 feet (2,440 meters) in the Guatemalan mountains, where clear nights from early November to early April were frosty, another year between 5,000 and 6,000 feet (1,525–1,830 meters) on a Costa Rican mountain, below the frost line but so exposed to prolonged storms of rain and chilling mist that it was often difficult to keep warm, and on shorter visits to other tropical heights. With a single exception (the Blue-throated Green Motmot of which I shall presently tell), the only birds that I found lodging in dormitories belonged to families that occupy them at low as well as high altitudes; birds did not respond to the cooler climate of the highlands by providing dormitories for themselves.

Nevertheless, it may be true that certain birds that range from warm lowlands to high tropical mountains, such as Southern House-Wrens that sleep in holes and crannies, and Banded-backed Wrens that build bulky closed nests with a side entrance, are able to reside over an altitudinal range of over 9,000 feet (2,740 meters) thanks to their sheltered sleeping places. The localities where I gave attention to this problem were all heavily forested. In the high Andes, above timberline, where the climate is much more severe, the situation is different. Similarly, although few birds of middle and high northern latitudes construct dormitories, harsh weather, especially when food is scarce, often causes them to crowd into whatever enclosed spaces they can find. We shall return to this matter in Chapter 11.

The dormitory habit runs in families and is closely associated with the forms of their breeding nests. In the New World, the chief dormitory-users are woodpeckers, woodcreepers, ovenbirds, wrens, and nuthatches. Woodcreepers and ovenbirds are confined to the Neotropical region, which includes all of South America and tropical North America; wrens are widespread in the New World, chiefly in the tropics, with one species extending to Eurasia; woodpeckers and nuthatches are more cosmopolitan. Woodpeckers carve holes in trees for breeding and sleeping. Woodcreepers nest and sleep in cavities that they find ready-made in trees and decaying stubs; ovenbirds construct an amazing variety of nests or breed in holes in trees or burrows in the ground. Wrens build covered nests amid vegetation or open nests in diverse nooks and crannies, also in birdhouses. Nuthatches nest in holes in trees, crevices in cliffs, or birdhouses. In the

Old World the principal dormitory-users, as far as now known, are weavers and Old World sparrows, who construct covered nests of many types or breed in boxes or crannies. Wherever one encounters birds of these families, one may expect to find them sleeping in nests or other enclosed spaces rather than in the open, although exceptions occur. In certain other families, dormitories are absent or rare, and usually associated with a nest or nest site not typical of the family.

Birds who sleep in dormitories, whether holes that they have carved, crannies in buildings, or nests that they have built, nearly always keep them clean. Either they do not defecate in their bedrooms or they remove their wastes. Some hole-nesting birds, such as trogons, motmots, kingfishers, and jacamars, give no attention to sanitation and permit their nests to become filthy; they do not sleep in them. The behavior of other hole-nesters depends upon whether they will continue to sleep in the cavity after their young fly. Woodpeckers, toucans, and woodcreepers carry out wastes as long as they enter their holes to feed the nestlings. After the young can climb up to receive their meals at the doorway, making it unnecessary for the parents to enter, those that will not sleep in the hole after the fledglings' departure often neglect to remove droppings, whereas species that will lodge in it with their flying young continue to keep it clean, often making special trips to carry away heaping billfuls of waste. In contrast to dormitory-users, roosting birds are typically careless of cleanliness, often burdening crowded roosts with so much excrement that it kills the supporting plants or, in cities, fouls buildings, sidewalks, and passers-by.

How did the dormitory habit evolve? Might it not be that birds first built covered nests for sleeping, or carved holes in trees for the same purpose, and later discovered that these structures or cavities served well for eggs and nestlings? The facts that many birds breed in covered nests or holes but do not use them as dormitories, and that it is rare to find a bird habitually sleeping in a situation more sheltered than that in which it rears its families, weigh against this supposition. Apparently, birds provided roofed or closed receptacles for their eggs before they slept in such structures. Tropical birds may sleep in dormitory nests more frequently than those of higher northern latitudes because they built roofed structures primarily to protect eggs, young, and brooding parents from intense tropical sunshine and torrential tropical downpours, and they afterward found that these structures make comfortable dormitories. Further evidence for the priority of the breeding nest is the fact that some birds who lodge in it after their young have flown fail to provide dormitories for themselves after this nest is lost or becomes too dilapidated to shelter them.

Starting with this primitive situation, we may trace the probable evolution of the dormitory habit through the following stages:

A. The breeding nest is used as a dormitory while it lasts; none is made or acquired solely for sleeping.
   1. The female sleeps in her nest when it contains neither eggs nor young.
   2. Both parents sleep in the breeding nest, sometimes with their helpers.
   3. Fledglings return to sleep in the nest, alone or with one or both parents, and sometimes also with helpers.

B. Nests or enclosed spaces similar to that in which the bird breeds are made or acquired for use as dormitories throughout the year.
   4. Self-supporting individuals sleep singly in nests or cavities at all seasons.
   5. Pairs lodge together in nests at all seasons.
   6. Fledglings and young recently self-supporting sleep in the nest with one or both parents.
   7. Long after they are self-supporting, young sleep in nests or cavities with their parents.
   8. Larger groups occupy dormitories.

In the following chapters, we shall consider dormitories in this sequence. Although present information does not enable us to trace the whole course of development from stage one to stage seven or eight in a single avian family, certain families, particularly woodpeckers and wrens, exemplify large segments of it, and numerous examples of each of these stages are available among birds as a whole.

Only in species included under B is the dormitory habit fully developed. Birds that use only their breeding nests as dormitories may not have the good fortune to retain them until the next breeding season; the nest or cavity may fall, decay in wet weather, or be taken from the original occupant by some other bird, or by a mammal. The bird who does not then build or acquire a shelter specially for sleeping roosts in the open.

References: Fraga 1980; Skutch 1961.

SIX

The Breeding Nest as the Only Dormitory

Hanging from slender twigs and dangling vines beside shady country lanes over much of tropical America are nests so conspicuous and unusual that they attract the attention of wayfarers with little interest in birds. Others are found in similarly exposed situations in shady pastures, gardens, and plantations with scattered trees, or at the forest's edge. Seven or eight inches (18–20 centimeters) long, these nests are shaped like a chemist's retort, with an entrance tube, opening downward, leading up to a rounded chamber that is attached at the top to the supporting twig. Composed of fine fibrous rootlets, the lustrous fungal filaments aptly called "vegetable horsehair," and similar materials, all closely matted and bound together with cobweb, the nest is nearly black. It is much more likely to be noticed than its builder, a small olive-green and pale yellow flycatcher, one of the many equally plain species that puzzle birdwatchers. Variously known as the Yellow-olive Flycatcher or Sulphury Flatbill, it ranges from southern Mexico to northern Argentina.

Without help from her mate, the female flycatcher builds her nest in two or three weeks. As it nears completion, she sleeps alone in it, often for as much as seven to ten nights before she lays the first of her two to four eggs. She continues to occupy it by night while for seventeen or eighteen days she incubates without help from her partner, and during the twenty-two to twenty-four days while both parents feed the nestlings. After their first flight, the young do not return to their pendent cradle, but their mother continues to enter it in the evening. If all goes well, she may continue to use her breeding nest as a dormitory for as long as four months; but usually the nest, wetted by almost daily rain, does not retain its shape so long, or it is claimed by some other bird for nesting or sleeping, or wasps oc-

Nest of Yellow-olive Flycatcher

cupy it. After the loss of her dormitory, the female Yellow-olive Flycatcher does not build another to shelter herself from the year's heaviest rains but roosts amid foliage, as does her mate at all seasons. Although I have never found one of these flycatchers sleeping otherwise than in a nest, I make this statement with confidence, for after the breeding season I have never seen one of these birds building, or found a new nest, around our house where these unmistakable structures hang prominently. The Yellow-olive Flycatcher exemplifies what is apparently the earliest stage in the development of the dormitory habit.

A neighbor of the Yellow-olive Flycatcher is the Sulphur-rumped Flycatcher, an olive and pale yellow bird who is more at home in humid forests but meets the other at its edges. Her nest tapers from a rounded bottom to a pointed apex, where it is attached to a slender twig, dangling vine, or hanging root of an epiphyte. In its side is a wide doorway shielded by an apronlike projection, continuous with the outer walls. The solitary female who builds and attends this nest may sleep in it for a few nights before she lays her two eggs and after her young abandon it, never to return. Her tendency to occupy her nest as a dormitory is less developed than that of the Yellow-olive Flycatcher.

For several nights before she lays her first egg, the female Eastern Phoebe, another flycatcher, sometimes sleeps in her newly finished half-cup of mud, rootlets, and fibers, beneath a bridge, in a building, on a cliff, or in some other sheltered spot. In late summer, after the female and her brood have departed, the male phoebe may sleep in the abandoned nest. Phoebes like to roost in sheltered places. I have found Black Phoebes sleeping on electric wires on a porch and beneath eaves in a narrow space between two buildings.

Even when they usually roost amid dense foliage, female titmice frequently sleep in their nest cavities before they start to lay; a Plain Titmouse did so for a month. Female Great Tits use their nest boxes as dormitories throughout the breeding season, when their mates roost amid foliage. In rocky areas amid the forests of West Africa lives the bald White-necked Rockfowl, a bird so strange that it is sometimes placed in a family with only one other species, although some ornithologists classify it with the babbling thrushes. In small colonies, rockfowls attach open nests of mud with fibrous binding, lined with vegetable materials, to the walls and roofs of caves, to sheltered spots on cliffs, or beneath the overhang of huge boulders. The birds sleep, apparently always singly, on empty nests that have been or will be used for breeding. Probably both sexes sleep on nests for lack of any other place to rest on nearly vertical rock surfaces.

After losing eggs or young, Yellow-billed Tropicbirds of Ascension Island return periodically to the island for three or four weeks and attempt to sleep in their nest cranny. If they find this occupied by another pair, they may rest in nearby sites, sometimes turning out a chick or attacking an incubating adult. Unoccupied tropicbirds prefer to sleep in established or potential nest sites rather than on open ground. On islands free of terrestrial predators, a booby often passes the night beside its incubating or brooding mate. The nocturnal Elf Owl of the arid southwestern United States and Mexico may repose in the nest cavity by day for a week or two before she lays an egg.

Nearly always the parent bird who incubates and broods is alone through the night, with his or her mate at a distance, a separation that reduces the chances that both will fall victims to the same predator. However, exceptions to this rule occur, especially when the nest is well enclosed and/or relatively safe. As soon as a pair of Orange-fronted Parakeets of the Pacific slope of Mexico and Central America have finished the excavation of their ample chamber in the heart of a large, black termitary, both sleep in it, as they continue to do throughout the periods of laying, incubating, and rearing the young. Only the female incubates. When not occupied with their nests, these parakeets of semiarid regions roost in small flocks in trees and arborescent cacti.

Pygmy parrots of New Guinea and neighboring islands also nest and sleep in cavities that they excavate in arboreal termitaries. One such chamber housed six adult Yellow-capped Pygmy Parrots and two nestlings of different ages. Even when not breeding, Buff-faced Pygmy Parrots sleep in their holes in termites' nests. Eight adults, who occupied a chamber without eggs or nestlings, remained within until well after sunrise and returned to their dormitory early in the evening. On rainy or cloudy mornings, they lingered inside until eight or nine o'clock. Probably pygmy parrots represent a more advanced stage of the dormitory habit, but in the absence of more information I provisionally place them here.

Among swifts, both parents build, incubate the eggs, and brood the nestlings, and frequently both sleep in, on, or close beside nests that are usually in sites difficult to reach. Before they lay an egg, during incubation, and while attending nestlings, both sexes of the Common Swift of Europe sleep in the nest space, in a tower, cranny in a house, or cleft in a precipice. If caught in a shower, they promptly seek shelter in their nests. After their first flight, the young do not return, but the parents lodge there for a few more nights before they migrate to Africa. There are good reasons to believe that these swifts pass some nights wholly in the air.

White-rumped Swifts in Tanzania occupy the closed, mud-walled nests that Abyssinian Swallows attach to the sides of buildings, beneath the eaves. After the swallows have abandoned a nest, the swifts line the lower half with feathers and plant down, glued in place with so much saliva that the surface glistens. They may continue to use the same nest for twelve or more years. Both sexes sleep in the remodeled nest before and while the eggs are laid and incubated and during the five to seven weeks of the nestling period. On quitting the nest, the young promptly fly beyond sight and do not return. The parents continue to lodge in the nest throughout the breeding season, during which they may rear three broods in more than half the year. Where White-rumped Swifts sleep through the remainder of the year and how the fledglings spend the night seem not to be known.

Another African species with a white rump, the House Swift, attaches a hemispherical bag beneath eaves. The roof tiles form the nest's ceiling; its walls are of coarse dead grass and sometimes feathers glued together with the birds' saliva into a rough and untidy but strong fabric, with from one to three round entrance holes in the lower side of the bag. At times a tunnel up to 3 inches (7.5 centimeters) long leads into the chamber. Both members of the pair build this unusual home and both sleep in it, before laying the eggs, during the approximately twenty-three days of incubation, and until the nestlings depart. During the day, when one comes to replace its partner on the eggs, the pair may remain together in the nest for a short interval. After abandoning their snug nursery at the age of about thirty-eight days, the young swifts do not return to it.

In tropical America, Lesser Swallow-tailed Swifts build, of vegetable down and other fluffy materials, a long, sleevelike nest, entered from the lower end. Near the middle they add an inwardly projecting bracket for the reception of their eggs. The nest hangs beneath a stout limb high in a tree or in an angle of a building. Both parents lodge in it, beginning long before its completion and continuing while it contains eggs and young. The incubating or brooding parent rests on the bracket, while its partner clings to the inside of the tube beneath it. They do not sleep continuously through the night but sometimes work inside the nest long after dark. Because the young swallow-tailed swifts in the single nest studied by François Haverschmidt in Suriname fell from it prematurely, we do not know whether, had they lived to fly, they would have returned to lodge with their parents, who continued to sleep in it after their loss. These swifts promptly repaired a slit cut in the nest's wall to inspect the contents.

In the vast Niah Cave of Sarawak, North Borneo, three species of "edible-nest" swiftlets of the genus *Collocalia* attach their bracketlike nests, made with much saliva and other materials, to high walls. Even when they have neither eggs nor young, pairs sleep close together in or beside their nests. Others, probably nonbreeding subadults, likewise sleep in pairs at night, but not close to nests.

Among the most charming of nests are those of the Bushtits that inhabit the highlands of Guatemala and southern Mexico. The pensile, pear-shaped pouch, about 6 inches (15 centimeters) long, is everywhere covered with finely branched, gray foliaceous lichens. A thick cushion of plant down in the bottom is overlaid with a coverlet of downy feathers, upon which four white eggs rest. At the top, between the arms of the supporting fork, is a sideward-facing, hooded doorway. In bushy openings and pastures amid woods of oaks and pines, these downy pouches are suspended from shrubs, about 8 to 12 feet (2.4 to 3.7 meters) up. The male and female Bushtits take turns incubating, and both sleep in the nest that both have built. After the young hatch, from one to three volunteers, black-faced like the male parent, help to feed and, less often, to brood them. Most or all of these attendants sleep in the nest on chilly nights in the high mountains. One pouch was occupied by four nestlings, two parents, and two helpers. After the young fly at about eighteen days of age, the nest hangs deserted, while all the Bushtits roost in trees. The downy pouch absorbs too much water to be a comfortable or healthful lodging during the cold rains that begin about the time the young fledge. A similar situation is found in the northern part of the Bushtits' range in the western United States, where three or four adults may sleep in the brood nest, but after breeding the tiny birds roost in trees through cold winter nights.

The nest of the Long-tailed Tit of Eurasia is no less beautiful than that of the Bushtit. Oval in form, with a domed top and a narrow round doorway in the side, it is supported from below, in a hedge, thorny bush, or tree, rather than suspended from its apex. The walls, thickly felted with shredded wool, green moss, cobweb, and the like, are covered on the outside with lichens, like the Bushtit's nest. But instead of being lined chiefly with vegetable down, this northern nest is padded with downy feathers, sometimes two thousand or more packed into it in a miracle of compression. Yet at nightfall the male Long-tailed Tit somehow manages to find room for himself in the nest, along with his mate, a dozen babies more or less, and all those feathers. In this species, also, one or two extra adults often help the parents to feed the nestlings. I have found no statement as to how these helpers sleep; at least, we may suppose that they do not insist upon lodging in the nest, which cannot be indefinitely distensible. Like their faraway relatives, the Bushtits, Long-tailed Tits roost in the open after breeding. In winter they pass the night cuddled together in balls, in holes, in dense thickets, or sometimes in old nests of their own species or of Winter Wrens.

In the southeastern United States, both parent Brown-headed Nuthatches often sleep in a nest box or other cavity, beginning before

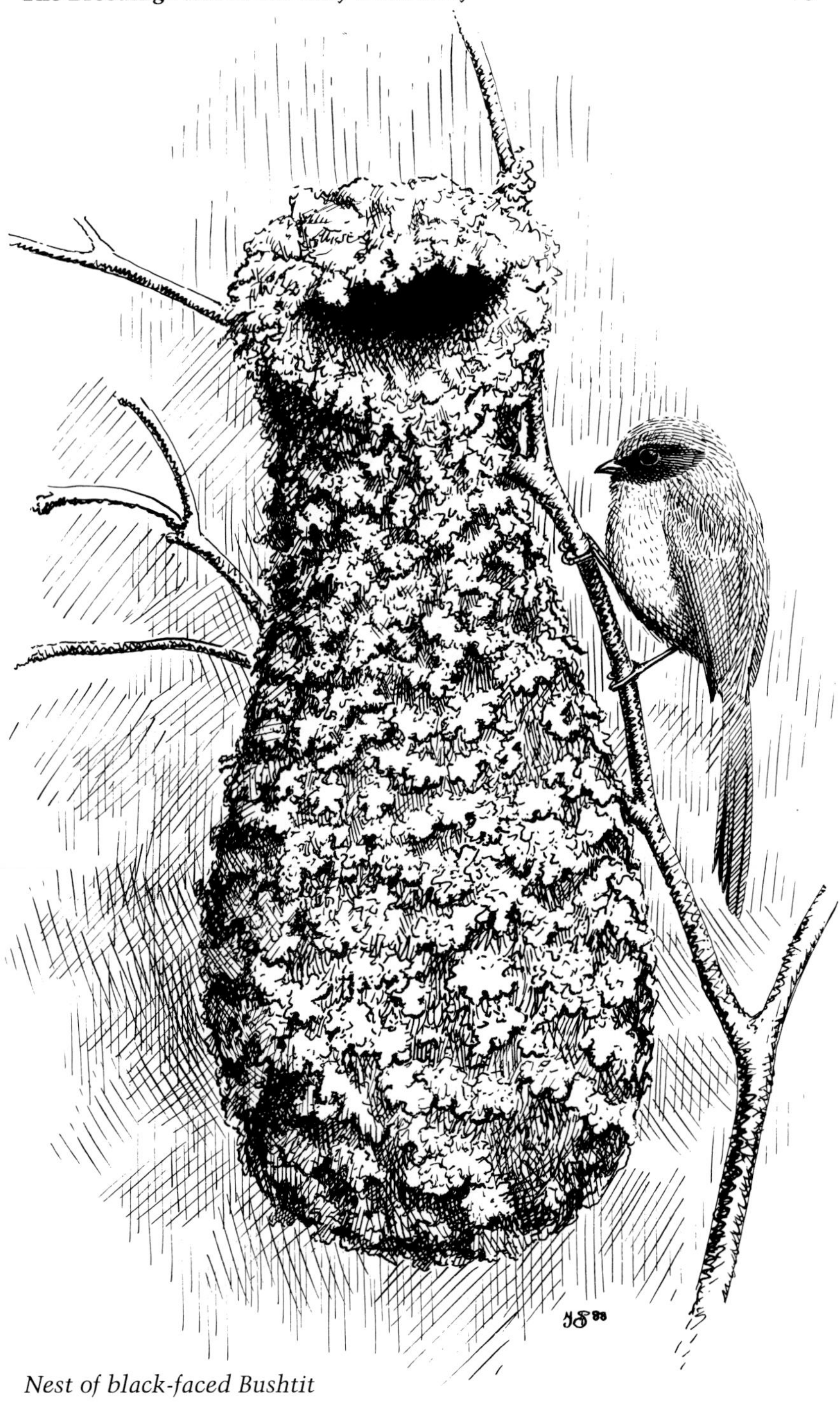

Nest of black-faced Bushtit

eggs are laid and continuing until the young are almost ready to fly. After the latter emerge, they and their parents roost amid vegetation rather than in an enclosed space. On a cold December night, however, four Brown-headed Nuthatches slept in a nest box, thereby approaching the Pygmy Nuthatches' more consistent occupation of dormitories.

The African Paradise Flycatcher builds its mossy open nest over water, amid green vegetation with which the green structure blends. In this fairly safe situation, the male, who takes a large share of incubation, brooding, and feeding the young, sleeps close beside the nest where his mate sits through the night. African Mountain Wagtails nest in holes in the downstream face of an old dam, only a few inches from the outrush of water, on the flanges of a girder beneath a bridge, or amid tangled vegetation on the trunk of a dead tree standing amid impounded water. Both sexes incubate the two eggs and attend the young. At night, both sleep on or beside the nest.

In Venezuela, I found four Tropical Mockingbirds, apparently parents with young as big as themselves, sleeping in a small, densely branched orange tree, a yard or so from one another and from the open cup of coarse twigs, where three well-grown nestlings were no longer brooded. Although this spiny orange tree was hardly a "dormitory," the roosting of grown birds so close to an occupied nest is so unusual, and so closely approaches the situation where the second parent sleeps on or beside an open nest, that it seems to deserve mention here.

Contrasting with the two white-rumped African swifts that prefer well-enclosed nests, the little gray Palm Swift is content with an amazingly slight structure. Using their saliva as glue, these birds attach to the more or less vertical surface of a hanging dead palm frond a loose pad of feathers mixed with plant floss. At its lower edge, this pad is extended outward in the form of a rimmed shelf or flange, narrower than the length of the eggs it will support. As soon as laid, each of the two eggs is glued by the female's saliva to this narrow shelf and the pad at its back. Here, standing on their narrow ends, the eggs rest, fixed immovably, through all the three weeks of incubation, without ever being turned.

Despite the slightness of their nest, both parent Palm Swifts sleep clinging to it in an upright, woodpeckerlike posture, beginning this habit when construction has barely begun, and continuing until after the young have flown. Often, before the eggs are laid, they rest in this position by day, the two partners touching each other, one sometimes with a wing extended protectingly over the back of its mate, and remaining so for periods of from a few minutes to over an hour. Both parents share the incubation of the eggs. The one on duty is often reluctant to leave when the other arrives to replace it. Then the newcomer may cling beside the first, sometimes nudging to hasten its departure, until it relinquishes the coveted post.

The nestling Palm Swifts remain, clinging upright to the pad of feathers and later at times crawling on brief excursions over the surrounding surface of the palm leaf, in much the posture of the incubating or brooding parent, for from twenty-nine to thirty-two days. Solitary nestlings usually fly a few days earlier than those in a brood of two. After its first flight, the young Palm Swift returns to its nest to sleep with its parents for only a night or two, and in some families not at all. Thus, the Palm Swift forms a transition between species whose fledglings do not return to sleep in the nest and those that do so more regularly, and for more nights.

In the United States and southern Canada east of the Rocky Mountains, Chimney Swifts build, inside a chimney or in some similar situation, a bracketlike nest of fine twigs glued together, and to its support, with their saliva. On or close beside this slight nest both parents sleep, clinging upright, while incubating their eggs and raising their nestlings, often with a helper or two close to them. After the fledglings' first flight, the whole family returns to sleep in the chimney for about a fortnight. Then, for another week or ten days, only two individuals, apparently the parents, enter the chimney for the night. Soon after this, Chimney Swifts gather in large flocks in preparation for their long flight to Peru and northwestern Brazil. While migrating, hundreds or thousands lodge for the night in a hollow tree or disused factory chimney, into which they funnel "like smoke flowing the wrong way." If these swifts can find enough chimneys or hollow trees in which to sleep in their wilder winter homes, they might be included among the yearlong dormitory-users.

Parent European Bee-eaters sometimes lead their fledglings back to the nesting colony, where the young birds retire for the night in the burrows where they hatched. The parents, however, fly off to their distant communal roosts, and the young often accompany them. In this migratory bird, the dormitory habit is not as well developed as in some of the resident African bee-eaters which we shall presently consider. Only one parent, usually the female, passes the night in the burrow with the eggs and nestlings.

Four young Pale-headed Jacamars returned to sleep with both parents in their deep natal burrow in a vertical exposure of clay, above a ravine in northern Venezuela. Here the whole family still lodged on my last visit, nearly two months after the first flight of the young jacamars, who now seemed well able to feed themselves. Possibly

Pale-headed Jacamar at nest burrow

these small, rather plain jacamars are yearlong dormitory users, but in the absence of more prolonged study I place them provisionally here. The better-known, more widespread Rufous-tailed Jacamar has not been found lodging in a burrow.

The cosmopolitan swallows sleep in diverse ways, some of which were considered in Chapter 4. They nest either in some enclosed space, such as a hole in a tree, a burrow in the ground, a niche in a cliff, a cave, a cranny in a house, or on a rafter in a barn, or they build for themselves a well-enclosed nest of clay or mud. When the nest is an open cup in a sheltered situation, one parent often sleeps upon the rim, with its breast above its incubating or brooding partner. When the nest is an urn of clay, both parents often lodge inside while raising their family. Pairs have been found sleeping together in or on the breeding nest in the Barn Swallow, Cliff Swallow, Wire-tailed Swallow of Africa, Bank Swallow or Sand Martin, Dusky Sand Martin, House Martin, Purple Martin, Black-capped Swallow, and Blue-and-white Swallow. Although the male Violet-green Swallow may not join his mate in a birdhouse, he may rest upon its roof through the night. A male Tree Swallow may begin the night in the same fashion, then fly off through the darkness to roost elsewhere—unusual behavior for a diurnal bird. Females of both the Northern Rough-winged Swallow and the Southern Rough-wing sleep alone with their eggs and nestlings, as does the female Rough-winged Bank Martin in Africa.

In some swallows whose males accompany their incubating or brooding partners through the night, their behavior is not consistent. Among the Bank Swallows studied in Minnesota by Arnold J. Petersen, both sexes were present in the nest burrow only 25 percent of the seventy-six times when nocturnal occupancy was determined. The female was alone 62 percent of the times and the male alone 10 percent. They were together slightly more frequently when the nest held eggs than when it contained nestlings.

An exceptionally thorough study of the sleeping arrangements of a swallow was made at a colony of Purple Martins in northern Texas by Charles R. Brown. In large martin houses, some males claimed a number of rooms. While pairs were forming, females were more eager to sleep with males than the reverse. When martins returned to the birdhouse in the twilight and a male with more than one room entered and remained briefly in several of them, a female often followed him. If he did not wish to share a room with her, he moved to another. If she persisted in joining him in this one, he exited again. This might continue for five minutes or more, until the female ceased to intrude into the male's chamber and they slept separately. Some exceptionally acquisitive male martins claimed eight or more compartments, creating a problem for a female who arrived after he had retired into one of them. She might peer into room after room until she found him. By sleeping with a male, a female indicated her choice of a partner and his territory, but males often rejected her first advances by refusing to lodge with her.

After the martins' pair bonds were cemented and nest-building began, partners slept together, in the room chosen for raising the family, on 80 percent of the nights; but to defend more compartments than they needed, some males slept alone. After laying her first egg, a female Purple Martin always slept with her nest. How long parents accompanied their nestlings at night depended upon the size of the brood and the date. All females brooded their young nightly until they were two weeks old. Females with five to seven nestlings over fifteen days old were less likely to sleep with them than were females with smaller broods, probably because the latter were less able to keep each other warm than were the young in more crowded nests. Males who slept with their mates while they incubated continued to accompany them for a time after the eggs hatched, but male parents of large broods might cease to pass the night with them even before they were two weeks old. Until mid-June, pairs who no longer slept with their young often lodged in another of the rooms they claimed. After this date, parent martins either continued to occupy chambers apart from their broods or they roosted in trees, often with members of one to six colonies together in a tree or grove within 0.6 mile (1 kilometer) from where they nested. These martins left their houses in the evening twilight and returned at daybreak to feed their young.

After the young martins flew from their nest at different times, their parents assembled the scattered brood on a convenient perch, often on wires or television aerials. Unable to recognize their own fledglings individually, they might accept and feed others of similar age. In late afternoon, the parents led their young back to the birdhouse but could not always conduct them to the correct room. Sometimes the young martins tried to enter chambers occupied by pairs still attending eggs or nestlings, or by House Sparrows, and were repulsed. Some fledglings did not find a room for sleeping until after nightfall. The number of nights that juveniles passed in the birdhouse after their first flight ranged from one to twelve, and was greater before than after mid-June, when Purple Martins began increasingly to roost in trees. A midday rainstorm might send a brood to its sheltering chamber, to remain until the shower stopped. Like migratory Purple Martins, Gray-breasted Martins, led by their mothers, return to their nests for from one or two to about ten days after their first flights.

Some fledgling Tree Swallows and Violet-green Swallows return to sleep in their nests; others do not. Young Bank Swallows, unable to distinguish their natal burrows from many others in a crowded colony in a sandbank, frequently enter neighboring tunnels, sometimes fledglings of three families in the same burrow. Fledgling Cliff Swallows usually return to their clay urns for only two or three nights. Northern Rough-winged Swallows scarcely ever return to their burrows, and Southern Rough-wings are not known to return. Juvenile Barn Swallows often continue to sleep close to their parents and sometimes help them to feed younger siblings of a second brood. Young House Martins sleep with both parents while the latter are engaged with second or third broods, which the juveniles help to feed. In Switzerland, ten or twelve martins were frequently found in a nest of hardened mud. Once parents, four flying young of each of the first two broods, and three nestlings of the third brood all crowded into a nest, raising the number of occupants to thirteen. In Costa Rica, Blue-and-white Swallows continue for weeks after they fledge to sleep in the nest space with their parents, and take shelter during daytime rains. These permanently resident swallows who occupy dormitories throughout the year will receive more attention in Chapter 10.

The tiny Cape Penduline Tits of southern Africa are among the very few birds that equip a closed nest with a hinged door. Their pendent pouch is composed of downy materials so well compacted that it sheds a heavy rain. From the side of the oval structure projects a short spout that looks like an entrance but is a sham, for its base is closed. Above this false doorway is a flap of material hinged at the bottom, closing the true entrance. To enter the nest, a tit alights on the spout, seizes the upper edge of the flap with its bill, and pulls it down. After climbing in, it turns around and closes the door. To leave, the tit pushes the flap out, exits, then nudges the flap upward until it sticks in place. In this ingenious structure the penduline tit deposits five or six eggs, which for additional security it buries in the nest lining during the laying period. Both parents sleep in the nest while it shelters eggs and nestlings.

Unlike Bushtits, fledgling Cape Penduline Tits return to sleep in the pouch with their parents, a habit which they continue while the adults rear a second brood. The fledglings of this later brood also lodge in the nest, with their parents and older siblings. The number of occupants may be swollen by other tits that join them. After the breeding season, up to eighteen Cape Penduline Tits have been found sleeping in a single pouch, with the doorway closed above them, even in February, South Africa's hottest month, when at night the temperature of the outside air remains at 80 to 90° F. (27–32° C.). No nests appear to be built specially for use as dormitories; after mid-May, when the old pouches used for breeding have deteriorated, the tits have no longer been found sleeping in nests. From May until breeding is resumed in the following October, all apparently roost in trees and shrubbery.

Among thrushes, fledglings of certain hole-nesters return to sleep in their nests, apparently without their parents. In Africa, a young Red-breasted Chat retired at dusk into the underground hole where it grew up. Juvenile Northern Wheatears enter burrows or crevices near that in which they hatched.

Fledgling Gray-headed Waxbills of East Africa return to sleep in the breeding nest with their parents for a few nights, after which they roost in a dense bush or small tree. The same appears to be true of the Crimson-cheeked Blue Waxbill of the same region.

In many other species, the young bird's return to its nest is related, not to the dormitory habit, but to its progressive separation

Nest of Cape Penduline Tit

from its nursery. In a number of albatrosses, herons, ibises, raptors, and other large birds that breed in open nests on the ground, trees, or cliffs, the young may crawl, hop, or flit beyond the nest for distances that increase as they become stronger, after each excursion returning to rest or sleep in it until they are ready to fly away. Exceptionally, a fledgling pigeon, hummingbird, American flycatcher, jay, finch, or other small bird returns to pass another night or two in its open nest, as it is most likely to do if another member of its brood continues to be fed there.

Many young birds succumb to predation or exposure during the days immediately following their departure from the nest. By returning to a safe nest for even a few nights longer, while they grow in strength and control of flight and increase their familiarity with their environment, they increase their chances of survival. Nocturnal shelter is especially important to them in stormy weather. During the first four or five days after their initial flight, when most return to sleep in the birdhouse, Purple Martins survive well. All members of twenty-six broods that Charles Brown kept under observation lived until they left the vicinity. The juvenile missing from each of six other broods may have been adopted by another family.

References: Bent 1942; Brown 1978; 1980; Combellack 1954; Conder 1956; Edson 1943; Forshaw and Cooper 1977; Gaston 1973; Grimes and Darku 1968; Hardy 1963; Hartley 1917; Haverschmidt 1958; Kluyver 1950; Lack 1956; Ligon 1968; Lunk 1962; Macgregor 1950; Mayhew 1958; Moreau 1939; 1940; 1941; 1942a; 1942b; 1949a; 1949b; Norris 1958; Odum 1941; Petersen 1955; Sherman 1952; Skead 1959; Skutch 1960; 1968; 1981; Stonehouse 1962; Summers-Smith 1958; Swift 1959; Van Someren 1956.

SEVEN

Yearlong Solitary Sleepers in Dormitories

The dormitory habit takes a step forward when, instead of using only the breeding nest as a dormitory and sleeping in more exposed situations when this deteriorates or is lost, the bird provides similar shelters for itself throughout the year. Among these more provident birds we can trace a sequence, similar to that which we have noticed in the less provident species, from those who always lodge alone, through those who sleep at all seasons in pairs, to parents who admit their dependent young into their bedrooms and, finally, those who admit their fully self-supporting offspring.

As humans discovered long ago, wood makes tight and durable shelters. But long before humans built houses, woodpeckers, the best equipped of avian wood-workers, learned to carve, in fairly sound wood, chambers to protect their eggs, their nestlings, and themselves. With the exception of a few that, in open country where suitable trees are lacking, dig burrows in the ground, and one species, the Rufous Woodpecker of southern Asia, that breeds in an occupied ants' nest, woodpeckers rear their young and sleep in holes that they excavate in trunks and branches, or sometimes in birdhouses. When carving their nest chambers, the sexes take turns at pecking into the wood and throwing out the chips. A typical woodpeckers' hole extends straight downward into the trunk from a round, sideward-facing doorway at the top. The immaculate, glossy white eggs rest upon wood particles on the unlined bottom. By day, the sexes incubate by turns; at night, the male alone (with a few exceptions that we shall presently notice) accompanies the eggs and nestlings, while his mate sleeps in another hole.

Whether mates remain together after raising their families or each goes its own way, most woodpeckers lodge singly in holes that at all seasons they carve for themselves when needed. The male wood-

Male Red-crowned Woodpecker repulsing fledgling

pecker, who tends to be the more domestic member of a pair, often takes care to have a sounder hole than the female, who may move into a somewhat dilapidated one that he has abandoned. Similarly, his dormitory is more likely to be chosen for eggs and nestlings. This general pattern is followed by the Red-crowned Woodpecker, Red-bellied Woodpecker, Hairy Woodpecker, Lineated Woodpecker, Black Woodpecker, Ivory-billed Woodpecker, Northern Flicker, and apparently many others. Surprisingly, even the Red-cockaded Woodpecker of the southeastern United States, a cooperative breeder whose holes carved into living pine trunks are surrounded by a barrier of sticky resin that holds snakes aloof, conforms to this pattern. These protected holes take long to prepare; a female Red-cockaded Woodpecker who lacks one in her own territory may in the evening fly afar to occupy, surreptitiously, a spare cavity in the territory of a neighboring group.

Most of the woodpeckers that have been well studied leave their newly fledged young outside, while the parents themselves retire to sleep, singly, in the nest or some other hole. On stormy evenings on a Costa Rican mountain, I watched young Hairy Woodpeckers settle down on trunks exposed to chilling rain, while their parents retired into their dormitories. When a fledgling Red-crowned Woodpecker in southern Costa Rica tried to join either parent in the adult's dormitory, it was firmly repulsed. When a parent, about to enter its hole for the night, found one of its offspring already ensconced within, it evicted the presumptuous youngster. Juveniles clung through the night to the outsides of trunks until each could find an unoccupied cavity of some sort or carve one for itself, as woodpeckers can do at an early age.

I have only once known two Red-crowned Woodpeckers beyond the nestling stage to sleep in the same hole. After a young female had been out of the nest for a week, sleeping in the open, a prolonged hard rain that continued until nightfall sent her to seek shelter in the hole where she was reared. Arriving at the doorway, she was repulsed by her father, who continued to lodge there. Apparently, she was desperately in need of shelter, for despite repeated rebuffs she forced her way in past his pecks, as on a fair evening she probably would not have done. Father and daughter slept together through the night. Thereafter, the persistent juvenile remained in possession of the cavity; her father found another dormitory. As we shall see, other woodpeckers take better care of their offspring.

Distributed through the woodlands of continental tropical America are sixty species of woodcreepers (family Dendrocolaptidae) whose brownish and rufous, often streaked or spotted, plumage blends well with the trunks up which they creep, searching for insects and spiders in crevices of the bark or the moss and lichens that cover it. They nest in holes made by decay or by woodpeckers in trunks and branches, where they lay two or three immaculate white eggs on a bed of bark, lichens, moss, or similar materials that they carry in. The few species whose ways of sleeping are known lodge throughout the year, always singly, in cavities like those in which they breed, or sometimes in a deep furrow in the side of a stump. They appear not to prepare their dormitories in any way. Often the bedroom is a hollow palm or other trunk with an open top through which rain enters, which suggests that concealment is more important to them than dryness. No distinguishing feature reveals where woodcreepers will sleep. They retire so late in the evening that their movements are difficult to follow in the deepening gloom, and they fly out in the early dawn, all of which makes their dormitories hard to find. The Tawny-winged Woodcreeper, or Dendrocincla, who nests without a mate, is a truculent bird that sometimes evicts woodpeckers as large as herself from the holes in which they have retired. One evening I watched a female of this woodcreeper install two fledglings in separate hollow trunks, then fly to the dead tree where for several years she slept and nested alone.

The eight species of treecreepers (family Climacteridae) of Australia and New Guinea superficially resemble the woodcreepers but are not closely related to them. Like woodcreepers, treecreepers forage by climbing up trunks and over branches, but without using their rounded tails for support, while with fairly long, slightly decurved bills they probe for insects and other small invertebrates in the bark and among mosses. In a cavity in a tree or log, lined with grasses, soft bark, plant down, or animal fur, they lay two or three white or flesh-tinted eggs spotted with reddish brown. Like woodcreepers, they sleep singly in holes and crannies, even species with helpers, although cooperatively breeding birds often lodge together. Brown Treecreepers and Red-browed Treecreepers, cooperative breeders, sleep in hollow dead trunks or branches, sometimes their old nest sites, often at great height. White-throated Treecreepers, who nest in unassisted pairs, prefer fissures or shallow cavities on the surfaces of trunks, or a nook at the forking of a branch, sometimes within 2 yards (2 meters) of the ground. Possibly because they sleep in more exposed situations, White-throated Treecreepers retire later in the evening than the other species.

In keeping with the great diversity of nests built by species of the ovenbird family, widespread in South America and tropical North America, their ways of sleeping are varied. The tiny Plain Xenops

Streaked-headed Woodcreeper at dormitory in tree

nests in a neat little cavity, resembling a miniature woodpecker's hole, which it carves for itself in soft wood, or occupies after the piculets who made it have left. It sleeps, always singly, in an old woodpecker's hole or other cavity in a tree. If a second bird, probably its mate, tries to join a solitary xenops in the evening, both fly away and only one returns. The Thorn-tailed Rayadito of far southern South America sleeps in deep, narrow cavities in trunks and beams like those in which it nests, but details are lacking.

Hanging conspicuously from a dangling vine or slender branch in an open space in the forest or at its edge, the nest of the Red-faced Spinetail of northwestern South America and southern Central America is a bulky globular structure composed largely of green moss, thin vines, and fibers, with a concealed doorway near the bottom. Similar nests are used both for breeding and for sleeping. As far as I have seen, Red-faced Spinetails always lodge alone. The Striped-crowned Spinetail of central South America builds, in trees and bushes at no great height, globular nests of spiny twigs, lined with downy materials and feathers. When used for breeding, the nest has a single round doorway in the side. When occupied as a dormitory, it has two entrances at opposite ends, permitting the occupant to escape through one if a predator pokes into the other. To convert the dormitory into a breeding nest, the spinetail closes one of the doorways—all of which reminds us of the behavior of a very different bird, the White-browed Sparrow-Weaver of Africa. I find no information as to how many Striped-crowned Spinetails sleep in one of these nests.

The small, brown and olive Spotted Barbtail climbs over mossy trunks in humid montane forests from Costa Rica to Bolivia. Its only known breeding nest was a massive structure of moss and rootlets, attached to the side of a log that bridged a woodland brook. A vertical shaft, entered from below, led upward to a shallow depression in the moss where two white eggs lay. In the Coastal Range of Venezuela, Paul Schwartz showed me similar but less bulky structures, with the platform near the top too slight to hold eggs, fastened to the rough, vertical face of a cliff beside a highway. Here he found barbtails sleeping, always singly, throughout the year. As will be told beyond, the dormitories of some species of ovenbirds are occupied by families; but other species, such as the Rufous Horneros, sleep amid foliage rather than in their massive nests of clay, from which they could not escape if a predator blocked the doorway.

The Eye-ringed Flatbill, a plain, unobtrusive, olive-green flycatcher with a notably broad bill and prominent whitish circles around its large, dark eyes, inhabits woodlands from southern Mex-

Plain Xenops in nest hole

ico to Panama. Its nests are pensile structures, in form much like those of the Yellow-olive Flycatcher, but bulkier, sometimes over a foot (30 centimeters) long, and composed of much coarser materials, including dead leaves and whole, small, living orchid plants. In these conspicuously hanging, retort-shaped structures, the female lays two pale reddish brown eggs, mottled with darker shades of reddish brown, and raises her family with no attendant male. She may continue to sleep in a nest from which her young have just flown, but without them. At all seasons I have found these unsociable flycatchers lodging alone in nests similar to those in which they breed, but often less carefully made, with a shorter entrance-spout or scarcely any, and a chamber too shallow to hold eggs. Because the sexes are alike, I have not learned whether both sexes or only females build and occupy these dormitory nests. The Olivaceous Flatbill of Panama and much of tropical South America sleeps, also singly, in bulky nests of the same form.

Over much of southern Canada, the United States, and the highlands of Mexico as far south as Oaxaca, White-breasted Nuthatches reside permanently in deciduous, mixed, and, less frequently, coniferous woodlands. In old woodpecker holes, knotholes, cavities made by decay, or birdhouses, they lay from five to ten eggs, which both sexes incubate. Fledglings are not led to sleep in the nest or some other cavity, but cling upside down to the trunk of a tree beneath a sheltering branch. At least during the winter months, White-breasted Nuthatches sleep singly in holes in trees, often those carved by Downy Woodpeckers, which the two species may occupy alternately. Unlike woodpeckers, who prefer doorways just wide enough to pass through, a nuthatch may occupy a chamber with an entrance two or three times the size of its body, which may permit it to slip past the snout or paws of an animal, perhaps a raccoon, intent upon eating it. In contrast to most dormitory-users, who refrain from soiling their sleeping quarters, White-breasted Nuthatches defecate in them, then carry out their droppings as they leave in the morning—an unusual habit. As the breeding season approaches, a female may claim her partner's dormitory for her eggs, as woodpeckers sometimes do. In very cold weather, White-breasted Nuthatches depart from their solitary habit to huddle with many others in a sheltering cavity, as will be told in Chapter 11.

The sleeping arrangements of wrens are as diverse as their nests and social relations. Some sleep alone, others in family groups. Some lead their fledglings to a dormitory, others apparently do not. Among the latter is the Plain Wren of southern Mexico and Central America, in Costa Rica called the Chinchirigüí, an onomatopoetic rendering of its antiphonal song. An inhabitant of weedy fields, bushy roadsides, and scrubby pastures, this modestly attired brown wren builds nests of two kinds. That intended for eggs and nestlings is a substantial, roughly globular structure, with a side entrance shaded by a visorlike extension of the roof. Its walls are composed of blades and inflorescences of grasses and similar pieces; its bottom is lined with seed down and other soft materials. The dormitory nest is a flimsy, thin-walled, roughly cylindrical pocket with the long axis horizontal and the entrance at one end, placed amid low vegetation. Composed of grasses, tendrils, straws, and inflorescences with flowers still attached, it is unlined, and so loosely constructed that, if wakened by the beam of a flashlight, its occupant can push through the rear wall—which may be the reason why the wall is so weak.

I have never found more than one wren sleeping in any of these pockets, sometimes facing outward toward the doorway and at other times with its head at the rear. The sexes of wrens are alike, but twice I was sure that a male occupied the flimsy dormitory, while nearby was a substantial breeding nest in which his mate slept before she laid in it. At harvest time when the maize was drying, a Plain Wren slept upon a nodding ear, beneath a loose husk that, like a hood, sheltered it from cold September rains. The wren rested with its breast against the husk, its tail outward, its brown feathers fluffed out, revealing their gray bases, and its head buried out of sight in the ball of loose down. A few wispy grass inflorescences around the sleeping wren signified that this was its dormitory. Never having found more than one Plain Wren in any of a number of nests that I have visited by night, I am fairly certain that fledglings do not sleep with a parent, or two together. Once I found a juvenile lodging in an old nest of the Bananaquit, into which the larger wren did not quite fit.

From Costa Rica to Colombia, northern Venezuela, and the islands of Trinidad and Tobago, Rufous-breasted Wrens inhabit dense, vine-burdened thickets and forest edges. Throughout the year they live in pairs, which maintain contact by antiphonal singing while they forage for insects and spiders amid foliage where visibility is limited. Together a male and female build a thick-walled covered nest with a side entrance, from a few inches above ground to 40 feet (12 meters) up in a tree. Sometimes such a nest is used as a dormitory; but much more often I have found these capable builders sleeping, always singly, in covered nests made by other birds. This wren's dormitory may be an unfinished or deteriorating nest of a Riverside Wren; but in our garden I have more frequently discovered a Rufous-breasted Wren ensconced for the night in the nest of a

Bananaquit, sometimes with the builder nearby, protesting the intrusion. The same nest may be occupied alternately by these two species. One Rufous-breasted Wren slept in a snug niche on a broad basal leaf of a tank bromeliad, well sheltered by the rosette of broad, strap-shaped leaves above her. The mate of this wren slipped into a pocket in a brown, fluffy mass of the liverwort *Frullania* on the opposite side of the same Calabash tree. Another Rufous-breasted Wren passed the night in a niche amid these liverworts where a Tawny-bellied, or Spotted-crowned, Euphonia had formerly lodged. I do not know why a wren that builds, only to neglect, an excellent nest so often chooses to sleep in the flimsier structure of some other bird, or in a makeshift dormitory. I lack evidence that this wren installs its fledglings in a nest or some other shelter.

Few birds are so widely distributed and abundant over the length and breadth of tropical America, including its islands, as the tiny, yellow-breasted Bananaquit, which with a sharp, decurved black bill sucks nectar from a great variety of flowers and supplements its diet with minute insects and spiders gleaned from foliage. Avoiding the dark interior of closed forests, it frequents gardens, plantations, parks, flowering thickets, arid thorny scrub, and mangroves. The most indefatigable builder that I know, it spends many of its waking hours making nests for breeding and sleeping. Whatever the purpose, its nest is a well-enclosed structure with a round entrance facing obliquely downward. Situated in a shrub or tree, low or high, it is composed of coarse vegetable fibers, pieces of dry leaves, lengths of vines and weed stems, tendrils, papery bark, and more or less green moss. Nests intended for rearing a brood are built by both sexes and tend to be more substantial than those made for dormitories, often by the single Bananaquit who will sleep in them. However, I have detected no consistent difference between the nests; one built as a dormitory may later be used for eggs, and a breeding nest may become a dormitory.

Often I have watched Bananaquits enter their nests in the evening, the male sometimes after singing profusely; or with a flashlight I have peeped into these nests at night, to see a bright yellow breast filling the doorway. If the sleeper awakes, dark, shining eyes beneath broad white eye-stripes stare into the beam. Unless the nest is shaken or the visitor is needlessly noisy, the bird remains in its bedroom, to fall asleep again after the light is turned away. Never have I found two Bananaquits past the nestling stage in the same nest. If one tries to enter with its mate, it is firmly repulsed. Sometimes, when evening finds a Bananaquit without a dormitory, possibly because its nest has fallen or has been pulled apart by some bird gathering materials to build its own nest, the homeless one approaches the covered nest or dormitory of some other small bird, perhaps a flycatcher or a euphonia. If the owner is present, a tussle ensues, to end with either the rightful proprietor or the intruder remaining within.

Bananaquits lay two or three eggs, which for twelve or thirteen days the female incubates alone. Both parents feed the nestlings, by regurgitation, in the manner of those other nectar-drinkers, the hummingbirds. The mother often broods her young nightly until, aged seventeen to nineteen days, they leave the nest, well able to fly. She may continue to sleep in the nest, but the fledglings do not return to it. Apparently, they roost amid foliage until, while still in juvenile attire, with duller breasts and eye-stripes than adults, each quite competently builds a dormitory for itself. If this is situated in the territory of an older bird, the latter may dispossess the poor youngster, who must start again and again until it finds an unclaimed area, or can hold its own against an adult. Meanwhile, it roosts amid foliage, or sleeps in the open nest of a tanager or some other bird.

The orange, black, and white Troupial, Venezuela's national bird, is a handsome oriole whose habits diverge widely from those of typical members of its genus. Most orioles nest in long, pensile pouches or deep pockets that they skillfully weave for themselves; the Troupial builds no nest but captures a closed structure made by some other bird, in Venezuela commonly that of the Rufous-fronted, or Plain-fronted, Thornbird. Troupials have no difficulty finding these abundant nests, which hang conspicuously from trees or service poles beside busy highways, as well as from trees standing isolated in fields and pastures or scattered over the far-spreading *llanos*. Built by brown, wren-sized birds of the ovenbird family, these structures of tightly interlaced twigs are commonly much longer than broad. Nearly always they contain several chambers, one above another, each with its own entrance from the outside and no internal passageway connecting them.

A pair of Troupials who claimed a thornbirds' nest with five rooms began by pulling out sticks to enlarge the doorway of the lowest, for Troupials are much bigger than the nests' builders. In this lowest chamber a Troupial began to sleep a month before the first egg appeared, while the six resident thornbirds continued to lodge in an upper chamber. By day they worked on a new nest on the opposite side of the same small tree. On some evenings both Troupials entered, or tried to enter, their preferred lowest compartment, but no more than one ever remained. The bird whose subsequent behavior revealed it to be the male (the sexes are alike) slept for some nights in the room immediately above that occupied by his mate, then moved to the thornbirds' new nest, causing them to return to an upper chamber of their old nest, where the female now incubated in the lowest room. He continued to sleep in the neighboring nest while both fed the nestling Troupials and their mother brooded them by night.

After the young Troupials flew from the nest when twenty-one and twenty-three days old, they did not return, and I could not find where they went. After some days, an adult resumed sleeping in the compartment where they were reared. The mate of this bird prepared a bedroom for itself by enlarging the doorway of the lower chamber of a neighboring smaller nest belonging to a single pair of thornbirds, who continued to sleep in the higher of their two rooms. This pair of Troupials had invaded three thornbirds' nests, in all of which they at one time slept, always singly, and in one of which they raised nestlings. To accommodate themselves, they destroyed the eggs of the first family of thornbirds and the nestlings of the second. Fortunately for the thornbirds, in northern Venezuela they are much more abundant than their persecutors, the Troupials. The Troupial is the only member of its family, the Icteridae, known to occupy dormitories, and the only species of its genus, *Icterus,* known to pirate nests.

Widespread in tropical America, euphonias are very small, short-billed tanagers whose food is chiefly berries, above all those of the many species of the mistletoe family (Loranthaceae) that parasitize tropical trees. The males of many euphonias are glossy blue-black or deep violet above and often also on the throat, with yellow foreheads, or foreheads and crowns, and bright yellow underparts. Females are much more plainly attired in olive and yellowish. Euphonias are the only tanagers known to build covered nests with a side entrance, which they place amid a thick coat of moss on a tree, in crannies in decaying trunks and fence posts, in clusters of epiphytes, or in almost any site that offers lateral support for their cozy globes. They lay larger sets of eggs than other tropical tanagers and, unlike others, feed their young by regurgitation. As told in Chapter 4, other tanagers roost in trees, often with members of a pair close together but not in contact. The only euphonia about whose sleeping habits I have learned anything prefers more sheltered situations.

For a dozen years, until the trees grew old and fell, Tawny-bellied, or Spotted-crowned, Euphonias slept amid the profuse growth of liverworts, mosses, and larger epiphytes on some Calabash trees in front of our house. Most often the dormitory was a pocket amid a mass of the brown liverwort *Frullania,* but sometimes one slept among the clustered stems of a small orchid, or on the root of a woody epiphyte, stretched horizontally beneath a sheltering tuft of liverworts. On many nights two males and a female slept in these trees, each in its own nook, never two together. Often my flashlight's beam picked out the yellow breast of a male, or the tawny underparts of a female, filling the opening in a pocket amid the liverworts. Sometimes in the evening they retired after prolonged calling or singing; at dawn they darted out and flew swiftly beyond view. Although the euphonias scarcely altered the niches where they slept and added no materials, they lodged in sites much like those they choose for their nests, so that, without stretching a point, their pockets amid the liverworts might be called dormitories. Of all the tanagers whose ways of sleeping I know, Tawny-bellied Euphonias select the most sheltered situations.

African weavers, especially males, are tireless builders, and some of their nests serve them as dormitories. V. G. L. Van Someren learned that, in three noncolonial species that bred in his sanctuary in the highlands of Kenya, the Golden-crowned Weaver, Highland Spectacled Weaver, and Masked Red Weaver, the males slept in unlined nests near the retort-shaped structures in which their mates incubated and brooded nestlings. Colonial, polygamous Village Weavers crowd scores or hundreds of globular nests in a single tree, where they hang conspicuously because the birds pull away most of the foliage around them, probably to have a clearer view of approaching raptors. The male builds the shell of a nest and displays, hanging beneath the downwardly directed entrance, to attract a mate. She lines the structure with fine grass inflorescences and feathers if she accepts it for her two eggs, which she alone incubates for about two weeks. Both parents feed the nestlings during their nearly three weeks in the nest. In a single breeding season, a male Village Weaver may build up to twenty-three nests and attract seven mates, as many as five of them simultaneously. Because females accept only fresh, recently built nests, the male tears down his old ones and builds others. At one time he has only two or three unlined nests, in one of which he sleeps.

I have been unable to learn from published accounts how long male weavers occupy their dormitories and where females sleep when not nesting. Since some of these weavers breed twice in a year, they may lodge in nests through much or all of it, and, accordingly, I include them among yearlong solitary sleepers in dormitories. Some of the other birds in this chapter are also included on the basis of observations that do not cover the whole year. Those placed here with confidence, because their ways of sleeping were followed through a number of years, are certain of the woodpeckers, woodcreepers, and ovenbirds, the Eye-ringed Flatbill, wrens, Bananaquit, and Tawny-bellied Euphonia.

References: Collias and Collias 1964; Kilham 1971; Narosky, Fraga, and de la Peña 1983; Noske 198?; Oniki 1970; Skutch 1954; 1960; 1967; 1969a; 1969b; 1972; 1981; 1983b; Skutch and Gardner 1985; Stickel 1964; Van Someren 1956.

EIGHT

A Unique Sleeping Arrangement

The Blue-throated Green Motmot is in several ways a unique bird. It lives at much higher altitudes than any other of the nine species in the motmot family, which prefer low and middle altitudes; it lacks the racquet-shaped central tail feathers of most motmots; and it is the only one known to use its burrow as a dormitory. Clad almost wholly in a beautiful, soft shade of green, it has pale buffy cheeks, a small black patch over each ear, another on the center of its breast, and blue tips on its long central tail feathers. The green motmot is found among woods of oaks, alder, other broad-leaved trees, and pines in the highlands of Middle America from Chiapas in southern Mexico through Guatemala to El Salvador and Honduras, chiefly from 4,500 to 9,500 feet (1,370 to 2,900 meters) above sea level.

In the high mountains of Guatemala, in February, I found a number of burrows in roadside banks and steep, bare slopes on the sides of washouts. Two parallel ruts or furrows in the mouths of most of these tunnels were so fresh that I had no doubt that they were occupied by birds who shuffled in and out, but to identify them was not easy. Not only were they extremely shy, but they entered their burrows late in the evening and left in the early dawn, when the light was too dim to reveal their colors. Finally, I learned that if I stood quietly beside a burrow as night ended, I would hear low, musical murmurs emerging from the bank at my side. Sometimes these soft notes were repeated again and again; but usually, after first hearing them, I had not long to wait before a dim, shadowy figure shot out of the earth and promptly disappeared into the thicket below the road. Almost immediately, or after a few minutes, the first bird was followed by a second, passing so close to me that I heard the rustle of its beating wings. Then, from amid the dusky foliage, came a de-

Blue-throated Green Motmot in front of burrow in bank

lightfully mellow piping, full, soft, and clear, as the two partners joined their voices to greet the new day, while the brightest stars were fading from the clear sky of the dry season, above fields and pastures white with frost.

Each of nine burrows that I investigated was occupied nightly by a pair of Blue-throated Green Motmots, who were absent all day. I expected that, as the breeding season approached, these birds would dig new tunnels for their eggs. When none appeared, I prepared four of the old ones for study, by digging down from the top of the bank and making, at the inner end of each burrow, a small opening which I carefully closed after each inspection. I found the burrows to be 56 to 70 inches (142 to 178 centimeters) long and very crooked, with sharp turns to the right or left, where the birds who dug them had met a root or other obstacle. At the inner end, each tunnel widened into a vaulted chamber, from 10 to 14 inches long, about 8 inches wide, and 4 or 5 inches high at the center (25–36 by 20 by 10–13 centimeters).

No bedding of any kind had been carried into these rooms where the motmots slept and would later lay their eggs. The floor was covered by a great mass of the indigestible parts of insects, chiefly beetles, mixed with loose soil and a few seeds. These fragments, regurgitated by the motmots during many nights, revealed what they had eaten. The volume of regurgitated shards and exoskeletons left no doubt that the burrows had been occupied for a long while, for below the loose debris they were compacted in a hard floor of considerable depth. Otherwise, the dormitories were clean, with no traces of droppings and little odor. Another indication of long occupancy was the fact that not one of the burrows was abandoned as a result of my alterations, which sometimes were necessarily rather extensive, although I tried to reduce them to a minimum. When I prepared openings at the backs of straighter, less confusing burrows of lowland motmots and kingfishers, they were always deserted unless incubation was well advanced or nestlings were present. But long residence had made the green motmots tenacious of their dormitories.

In early April, just after the last of the nocturnal frosts, each of the four burrows that I had opened received, on alternate days, three plain white eggs, almost equally blunt on the two ends. They lay on the hard floors of the unlined chambers. Throughout the incubation period of twenty-one or twenty-two days, the parents continued their habit of sleeping together in the burrows. Whether one incubated all the eggs at night, or one parent covered two and the other sat upon one egg, I could not learn. Both flew from the tunnel at dawn, as they had long been doing, and in the neighboring thicket sang a melodious duet, but no more prolonged, and often briefer, than on the frosty mornings of February and March. Then, from three-quarters of an hour to a full hour, the eggs were alone. Soon after six o'clock one parent returned, to incubate until it was relieved by its partner about four hours later. The second bird sat through the middle of the day, after which the first returned to take charge in the late afternoon. At about six o'clock in the evening, this motmot flew out, leaving the eggs unattended while it sought supper. As daylight faded, both returned to the burrow, the second after the evening twilight had become very dim.

Although both parents had slept in the burrow with the eggs during the four nights that intervened between the laying of the first and the last of the set of three, they apparently applied little heat to them, for in two burrows all hatched within twenty-four hours, and in each of two other burrows two eggs hatched on one day and the third egg on the following day. The hatchlings bore no trace of down on their pink skins. A black protuberance on each side of the head showed where their eyes were hidden. Before they were a day old, they could stand erect on the full feet, with the swollen abdomen as the third point of the tripod. They could even walk a little in a halting, tottering fashion.

On the day their nestlings hatched, the parents, all shyness overruled by ardent parental devotion, remained steadfastly covering them while, not without noise, I opened the burrow. Reaching into a dark chamber, I took hold of a guardian bird—father or mother, I could not tell which—and slowly, carefully lifted it out, finding its feathers as fair to the eye as they were soft to the touch. After a brief, gentle effort to escape, the motmot silently, resignedly looked up at its captor with deep brown eyes, as soft in cast as its plumage was in texture. After I had examined and made notes on the nestlings, I lowered them into the burrow and replaced the parent over them, where it remained while I closed the hole, packed the earth over it, and departed. On hatching day, a parent stayed in three of the four burrows that I opened. From this point onward, parental steadfastness waned. After the nestlings were two days old, the adult brooding them invariably retreated beyond reach in the tunnel, or flew out, when I inspected a burrow.

Both parents fed the nestlings, chiefly with big, hairless caterpillars and other insect larvae, with an admixture of winged insects. Through most of the month that the young remained in the burrow, the adults followed their old routine of sleeping in it. A pair with only two offspring continued to accompany them at night as long as they remained underground. At another burrow, only one parent slept with the three nestlings on their last few nights in it. At a third burrow, first one, then the other parent ceased to enter at nightfall, leaving the nestlings, now well feathered and no longer in need of brooding, alone during the last four nights before they flew. Unless the parents who went elsewhere to sleep had shelters unknown to me, they must have found the change from their snug, subterranean bedrooms to wet, dripping foliage far from pleasant, for the wet season had now returned in full force, bringing cold nocturnal rains. But at least these parents escaped the importunities of their well-grown young, who in the earliest dawn set up an amazing din, clamoring for their breakfasts with loud trills and mellow calls, which might have been pleasant to hear if the youngsters had not all vociferated at once, with no attempt to keep time, in so confined a space.

At the beginning of June, the last of the young motmots flew from its burrow. I fully expected that their parents would carefully lead them to shelter from the inclement weather of this season, probably in their natal burrows. But not one of the eleven juveniles from the nests I had watched returned to its tunnel; and if any of them slept in some other sheltered place, I could not find it. Probably the young motmots roosted amid dripping foliage.

After their broods flew, the behavior of the parents varied from pair to pair. One couple slept in the burrow that their two young had just left, which they continued to occupy until the year's end. The parents who had left their nestlings quite alone during their final nights in the burrow never returned to it but apparently braved chilling nocturnal rains, as their offspring did. After the young left the burrow where a single parent had accompanied them during their last few nights in it, the other parent resumed sleeping there, and the united couple continued to occupy it until their new burrow was ready. Another pair of motmots, who also continued to lodge in their burrow after their fledglings flew away, abandoned it when, a few days later, a pair of Black-capped Swallows carried in leaves and pine needles for a nest.

Motmots and their relatives, the kingfishers, fail to clean their nests. During the month that the young Blue-throated Green Motmots occupied them, the burrows became foul with droppings. The parents awaited a favorable time to prepare new lodgings for themselves. At first they were probably too busy feeding their fledglings, whom I never saw, and the soil was too sodden with frequent rains for comfortable digging. But at the end of June, when the young had been in the open for about four weeks and could probably find much food for themselves, a lull in the rains made the earth more tractable for working, neither saturated and muddy nor dry and powdery.

Five pairs began their new burrows at distances ranging from 20 inches to 28 feet (0.5 to 8.5 meters) from those in which they had nested. In early July, I passed many hours in my blind, watching a pair of green motmots at work. Their plumage was worn and faded, for they had not yet molted to replace feathers frayed by their arduous labors of the preceding months. They had two periods of work daily, in the morning from about seven to nine or ten o'clock, and in the afternoon between three and six. Sharing the task, the partners labored in alternate spells of about three to twelve minutes. As each entered the tunnel, it kicked vigorously backward, throwing outward two parallel, intermittent jets of loose earth that followed the digger inward until they disappeared into the darkness of the shaft. The motmots never pushed or kicked the loose soil before them as they emerged after finishing a stint. While one member of the pair toiled inside, the other rested low in neighboring bushes, repeating over and over a low, soft note and at intervals swinging its tail from side to side in typical motmot fashion. Nearly always, the waiting partner entered as soon as its coworker flew out.

When the new burrows were ready, the motmots began to sleep in them. Single-brooded like most of the birds on this high mountain, they would not lay again until the following year. On many a fog-drenched dawn through the remainder of the wet season from July to October, and on frosty mornings of drier November and December, I stood in the dim light beside a burrow to count the motmots as they emerged. Nearly always two flew out. Twice I found three sleeping together, and twice only one occupied a burrow. But these arrangements were temporary, probably caused by the loss of a mate and solitary individuals seeking company at night. The motmots were now much less vocal than they had been in February and March. Standing beside a burrow, I rarely heard the soft, confidential murmurs which then had preluded their departure. After emerging, they sometimes sang a little, but on many a blustery November morning they were silent.

Most of the burrows were so crooked that from the front I could not look into the chambers where the motmots slept. But one evening in November, when darkness overtook me as I was passing through the forest far from my abode, it occurred to me to peep into a burrow that I had earlier found. To my surprise and delight, my flashlight revealed that it was so straight that I could see right to the end. There, in the center of the chamber, was a single mass of green, fluffy feathers, with no distinguishable features except a few largely concealed wing plumes. I could not decide how many birds were present until I roused them with repeated low sounds, when two unburied their heads from the downy mass, and one started to preen. I turned off the light, waited a minute, then peered in once more. The heads had already vanished into the single mound of plumage. The motmots must have felt very secure in their deep retreat, not to have been alarmed by the unprecedented intrusion.

Since Blue-throated Green Motmots lay their eggs in burrows where they have already lodged, nearly always in pairs, for nine or ten months, we might conclude that, reversing the usual order, in this species the dormitory becomes the breeding nest, and that, in the evolutionary sequence, the use of burrows for sleeping preceded their use for breeding. Comparison of this motmot with others makes this improbable. Burrow-nesters of several families prepare their tunnels months before they will lay in them, when the soil may be more favorable for digging. This practice has the further advantage that an aging burrow is less likely to attract attention than a new one with a pile of freshly dug earth below its mouth, with no accumulation of dead leaves and other litter covering it. Thus, in this valley in southern Costa Rica, Blue-diademed, or Blue-crowned, Motmots excavate chiefly from August to October, when the ground is moist and soft, burrows in which they will not lay eggs until the following March or April, when the soil may still be dry and hard. In this long interval, the burrow remains untenanted; as far as known, no lowland motmot occupies its burrow as a dormitory. In its process of adaptation to cold highlands where no other member of its family dwells, the Blue-throated Green Motmot became exceptionally forehanded in digging its nesting burrows, then started to sleep in them through the wettest and coldest months of a sometimes rigorous montane climate.

References: Skutch 1945b; 1983a.

NINE

Putting the Fledglings to Bed

In Chapter 6 we considered the return of fledglings to the breeding nests of parents who, failing to provide special dormitories for themselves, do not sleep in sheltered places throughout the year. In this chapter we shall give attention to birds who themselves occupy dormitories at all seasons, but sleep with or near their young only while the latter remain dependent, or a little longer. In the following chapter we shall look at birds with more enduring bonds between parents and offspring, who remain together for many months and make of their dormitory a family dwelling.

Our first example of the sixth stage in the evolution of sleeping arrangements is the little Blue-and-white Swallow. Of its two races, one nests in the South Temperate Zone in Argentina and Chile and migrates northward in the austral winter, rarely as far as southern Mexico. The other race breeds from northern Argentina to Costa Rica and is apparently resident throughout this vast area, where, at least in the north, it prefers middle and high altitudes. A very adaptable bird, both as to habitat and nest site, it builds its shallow cup of straws, grass blades, and bits of weed stems in situations as diverse as nooks in buildings, holes in trees, ready-made burrows in banks, crevices in masonry, and the like. Male and female share the task of building. The latter lays two to four eggs on a bed of downy feathers. Both sexes incubate for a period of fifteen days. At night one sits on the eggs or broods the nestlings while its partner rests on the nest's rim, its white breast upon the other's deep blue shoulder.

When, at the age of twenty-six or twenty-seven days, the young fly, they are led back to the nest space by their parents. On wet afternoons, they return early, to rest warm and dry, while their parents fly about in the rain catching insects for them. Each return of a food-bearing parent is greeted by a chorus of loud chirrups. Tired by so much flying under a wet sky, a pair that nested in my roof would at intervals rest for fifteen or twenty minutes beside their offspring, then fly out together to collect more insects in the rain. At night the family slept on the ridgeplate where they had nested, pressed close together in a row, the three young in the center and a parent at each end. When no longer dependent upon parental feeding, the juveniles dispersed; the last continued to lodge in the parents' nest space until a little over two months old. After the loss of a second brood, the parents slept side by side on the beam.

Another pair of Blue-and-white Swallows slept for two years under the roof tiles of my present dwelling, continuously except for two occasions when, the female having disappeared, the male was absent for a few days while he sought and brought back a bride. Each new partner had difficulty joining him in the confusing situation presented by many similar channels under the rows of tiles, with no connection between them. After entering his preferred space, the male would sing, over and over, while she flew around seeking the proper entrance at the edge of the roof, sometimes continuing until he emerged to guide her in.

The Southern House-Wren is a small, brown, insectivorous bird, so similar to the Northern House-Wren of North America that it is often classified as a race of this familiar species, although it differs in habits as well as measurements. A bird of wide ecological tolerance, it ranges from southern Mexico through Central and South America to Tierra del Fuego and the Falkland Islands, and from the coasts up to about 9,000 feet (2,750 meters) in the mountains. Except in the far south, it appears everywhere to be permanently resident.

Although Southern House-Wrens live in pairs throughout the year, the partners sleep separately in almost any nook or cranny that offers protection: a hole in a tree, a niche or short burrow in a bank (not made by the wrens themselves), in pockets amid the leaves of a thatched roof, beneath roof tiles, in a bunch of green bananas hanging where it grew, in the hollow end of a bamboo crosspiece of a garden trellis, in a birdhouse or a gourd prepared for them. The dormitory may be near the ground or so high in a branchless trunk that the wren does not fly directly to it but works its way by stages, clinging to the trunk after each short upward flight. In most of these situations it is difficult to see the sleeping bird, but in the highlands of Guatemala I found nine wrens passing the night in as many shallow niches and short tunnels in a roadside bank. The tenant of each niche slept with its head inward. All its body feathers were erected, revealing light, ordinarily concealed bases that made the little bird appear to have just flown into its shelter from a snowstorm—an illusion supported by the penetratingly cold mountain air. In each ball of downy feathers the sleeper's head was hidden; only its barred tail, projecting toward me, was clearly distinguishable. The tuft of feather-down pulsated with the rapid breathing of the tiny body in its midst.

In the tropical parts of its range, the Southern House-Wren has a much longer breeding season than most of its avian neighbors. Three or four broods may be reared in a reproductive effort that continues through most of the year. Nests are built in situations hardly less varied than those chosen for sleeping, although not in the highest of them, and include such unexpected places as the closed nest of some other bird, a saddle-bag hanging on a wall, or between the feet of a cylindrical rain gauge in a box on a post. If the chosen space is large, male and female set about with great energy, and much singing on his part, to reduce the volume with twiglets and other coarse pieces. In a depression in the top of this mass, the female builds the nest proper with fine grasses, vegetable fibers, rootlets, horsehairs, or pine needles where available, and lines it with downy feathers, fragments of snakeskin, often scraps of paper or cellophane. Here she lays, on consecutive days, from three to five, most often four, whitish eggs finely flecked all over with shades of brown. Beyond the tropics in Argentina and Chile, the number may be increased to six or seven, rarely more. Incubated by the female with no help from her songful and interested mate, the eggs hatch in fifteen days. Both parents nourish, with larval and mature insects, nestlings who spontaneously leave the cavity when eighteen or nineteen days old, well clad in plumage like that of their parents, and able to fly fairly well.

Toward the end of the fledgling Southern House-Wrens' first day in the open, the parents lead them to a sheltered place that they have already chosen for them. This may be the nest where they hatched and grew up; but if this situation is difficult for the young to reach, or if it becomes infested with vermin, the parents select a more accessible, or a cleaner, dormitory. The temporary lodging may be low, poorly closed, and used only for a few nights, until the fledglings fly well enough to reach a higher, safer, snugger bedroom.

Young wrens have a strong tendency to follow their parents, who, to lead them to a new spot, fly repeatedly from them to the chosen destination, until all have reached it. This same procedure is employed to guide fledgling wrens to a dormitory difficult to reach. To the accompaniment of the father's profuse singing, the parents fly again and again from the young brood to the entrance. They go in and out, in and out, until the weakly flying fledglings reach the desired objective, or until, darkness falling fast, they retire to their

Southern House-Wren fledglings re-entering nest for night

own dormitories, leaving the most backward member of the brood to fall asleep wherever drowsiness overtakes it. A night or two exposed to dew or rain appears not to harm it.

If the fledglings can enter their dormitory before the light is too dim, their parents often treat them like nestlings who have never been outside, bringing food to them and removing their droppings. Finally, as daylight wanes, their mother may join them for the night, as she is more likely to do if they can be put to bed in the nest where they hatched, or in some other well-enclosed space, than if, perforce, they are installed in a more exposed situation. Next morning, the young may linger in their shelter for half an hour after their mother has flown out, while breakfast is brought to them. As days pass, their times for retiring in the evening and becoming active in the morning approach those of the adults.

In the highlands of Guatemala, where Southern House-Wrens appeared to raise only one brood, in April and May when the great majority of birds of other kinds nested, I watched the gradual dispersal of a single brood of young wrens, without the complication of later broods. These wrens nested and slept in niches in a roadside bank, where they were easy to find. On the last day of May, four days after their departure from the nest, I found the four fledglings sleeping in a shallow niche in the earthen bank, about 90 yards (82 meters) from the similar but deeper cavity where they were reared. Their mother was not with them, probably because, without her, the cranny was filled to capacity by the closely packed young. Two nights later, the wrens lodged in another niche, 50 feet (15 meters) away. Screened by vegetation drooping from the top of the bank, this more capacious niche apparently contained the mother with her four offspring, all so tightly cuddled together that to count them was difficult.

After they had been out of the nest only nine days, the young wrens began to lodge apart. Three slept in one cranny in the bank, the other with a parent in a neighboring niche. Thirteen days after their exit from the nest, two juveniles slept together, one alone; the fourth had vanished. But at the end of June, a month after their first flight, I again found three wrens sleeping together, in the entrance to a burrow that a pair of Blue-throated Green Motmots was digging. They had chosen as their starting point one of the niches where the wrens lodged. As soon as they completed their subterranean chamber and started to sleep in it, the motmots scattered the wrens, who retired earlier and arose later than their bigger neighbors. After this, I found the wrens always sleeping in solitude. By the end of July, two months after the young wrens fledged, four, including both parents and offspring, slept in as many niches spread along 50 yards (46 meters) of the roadway. Four wrens, apparently always the same individuals, continued to lodge in the same stretch of bank where the nest had been until the end of November, and at least three of them until mid-December. Although all slept singly after the young had been out of the nest a little over a month, the latter remained near their parents for at least half a year longer.

In the valley of the Río Buena Vista in southern Costa Rica, I followed for two seasons the activities of a pair of wrens that reared seven broods in a gourd that I had tied in an orange tree in front of a window of my thatched cabin. Three fledglings of the first brood, hatched in April, were led by their parents to the gourd, where they continued to sleep beside their mother while she incubated her second set of eggs. After the nestlings of this second brood were born, their full-grown brothers and sisters were refused admittance by their mother when they came to join her in the evening. These juveniles were not easily turned away; but by the combined efforts of their mother, who pecked from within, and their father, who attacked from the rear while they clung in the doorway, they were finally persuaded to desist from their stubborn attempts to enter. Then the male parent slept with his mate and the nestlings in the gourd, where he was better able to resist the intrusion of the independent young—an unusual arrangement. After their eviction, the two-month-old wrens promptly vanished from the vicinity. When the fledglings of the second brood left the gourd, they were installed for a few nights in an old banana stump and other temporary quarters, until they were strong enough to fly up and enter pockets among the sugarcane leaves of my roof, where the parents now lodged. The young of the third brood were also led to sleep in the thatch above my head.

In the second year, the fledglings of the first brood were stronger and more precocious than their predecessors in the gourd, while the parents seemed to be more indulgent, a situation which enabled the young wrens to sleep with their mother while her second brood grew up. Two of the juveniles, a male and a female, fed their younger siblings, with whom they were so closely associated in the gourd. The situation was now hardly different from that of the cooperatively breeding birds whose sleeping arrangements will receive our attention in the following chapter. However, house-wrens have not made all the psychic adjustments indispensable for cooperative breeding; the young female became insubordinate and tried to displace her mother from the gourd. This led to a day of the fiercest fighting, between mother and daughter, that I have ever witnessed among birds of any kind. The father and brother watched the fray without participating in it. The unfilial daughter was defeated and soon disappeared. The less aggressive brother remained and continued to feed the surviving nestling, a male who, in due course, helped to nourish the third brood.

Years later, I had a nest box in which one juvenile of the first brood managed, despite parental opposition, to sleep with its mother until she hatched the second brood. The stage was now set for a repetition of the drama that I had earlier witnessed at the gourd; but, alas! some predator raided the box two days after the second brood hatched. I have never again seen a young helper among house-wrens.

Southern House-Wrens do not invariably lead their fledglings to covered shelters. For several years, those around my present abode in Costa Rica installed their newly emerged young in open, cup-shaped nests amid clustered leaves of Calabash trees in front of the house, or in high crotches amid the foliage where no nest was visible. Here the juveniles were mostly well screened by the thick leaves but hardly sheltered from the frequent nocturnal rains. In one case, the reason for this departure from the usual practice of house-wrens was apparently the invasion of their nest box by big, black, carrion-eating ants. Some of these young wrens were content for months to sleep in the open nests or leafy crotches where they were first installed by their parents—an interesting example of the persistence of a habit, acquired under parental guidance at an early age, contrary to the widespread practice of a species.

A neighbor of the Southern House-Wren in the highlands of southern Mexico and northern Central America, from Chiapas to Honduras, is the Rufous-browed Wren. More retiring than the house-wrens, it lives throughout the year in pairs in woods of pines and oaks, and in the cypress forests of high mountaintops, where its halting song is heard more often than the secretive bird is seen. In cavities low in trunks or niches in earthen banks, it builds cup-shaped nests of pine needles, grass blades, and slender stalks, lined with feathers, in which it lays three white, cinnamon-speckled eggs. Rufous-browed Wrens sleep in little tunnels in banks, old nests of the Banded-backed Wren, or high in great, moss-laden cypress trees. In November I found two couples, apparently mated pairs, sleeping together; but in March a wren that tried to join its partner in the evening was repulsed. Three fledglings and an adult slept in contact in a cranny in a roadside bank near the site of their nest, which a passerby had pulled from its niche.

In the mountains of Costa Rica and western Panama, the Rufous-browed Wren is represented by an allied species, the Ochraceous Wren. In the humid montane forests, segments of dead branches often break from high trees but are prevented from falling by stout roots of epiphytes that attach them to sounder wood. Swinging pendulumlike high in the air, from a few inches to more than a yard beneath their supports, thick lengths of rotting branches, embraced by matted roots, often bear a luxuriant growth of flowering shrubs and herbs, ferns, and mosses; they might be compared to the wire baskets in which house plants are often grown, hung high in a tree. Amid such massed aerial vegetation, Ochraceous Wrens nest. One evening, after fledglings had left one of these inaccessible nests, I watched two wrens, incessantly repeating their slow, plaintive *churr,* make their way laboriously up the tall trunk, then fly across to the dangling segment of branch and enter a cavity in it. About twenty minutes later, a third wren joined them, and all stayed until it grew dark, swayed through the night by every slightest breeze. Although the wrens were too high to distinguish young from old, I had little doubt that two young slept with a parent.

The Winter Wren (known simply as the Wren in Britain, where it is the only member of its family), spreads widely over the more northerly parts of the Northern Hemisphere, where, enduring rigorous winters, it resides permanently over much of its breeding range. The nest, a globe of moss and other vegetable materials, with a narrow round doorway in the side, is tucked into a cranny amid the roots and earth heaved up by a fallen tree or a niche in a wall, or it is built in a birdhouse, amid dense heather, inside a building, or in almost any nook that will hold it snugly. At times the abandoned nest of some other bird is used. The frequently polygamous male builds the shells of a number of nests; the female lines with feathers and fibers one that she accepts for her four to seven eggs, rarely more or less. She alone incubates for a period of fifteen or sixteen days.

At the age of thirteen to seventeen days, and sometimes more, young Winter Wrens sally from their nests and in the afternoon or evening are led by either of the parents, or both of them, to some snug shelter, usually not the nest in which they grew up but often one of their father's unoccupied nests, or at times to an old nest of some other bird, such as (in Britain) the Song Thrush, Greenfinch, Robin, Spotted Flycatcher, or Dunnock. The fledglings may even intrude into a nest of one of these species that contains the builder's eggs or nestlings. On their first days in the open, they are not very responsive to the parents' efforts to guide them to a dormitory but, impelled by their strong urge to cuddle together in a narrow space, they crowd into the first promising nest or nook that they encounter, or, on the other hand, ignoring parental attempts to move them onward, settle for the night in exposed, perilous spots.

More often than their mother, the father guides the fledgling Winter Wrens to their sleeping place, much in the manner of Southern House-Wrens. He advances toward the dormitory by stages, calling and singing as he goes. Next, he indicates the chosen nook by singing in front of it, often creeping in and out as he did earlier in the year when showing his nests to potential mates. When the female leads the young brood, she flies repeatedly between them and the dormitory, sometimes a dozen times. After they reach the nest or niche, she sings the dainty whisper or "swallow" song while she puts the fledglings to bed and after they are ensconced. If they are disturbed, the parent of either sex may lead them to another dormitory, and perhaps to a third if they are again alarmed, until in the failing light their guide leaves them where they happen to be. If the evening is not too far advanced when the young retire, their mother, and more rarely their father, may feed them in the dormitory, as though they were nestlings. The female Winter Wren may sleep in the dormitory with her fledglings, and occasionally she permits them to lodge in the nest where she incubates her second set of eggs. Rarely do juveniles sleep together for more than three or four weeks after they leave their natal nest. After separating, they often lodge singly in unoccupied nests of their own species, to which their attachment is strong.

In summer, adult Winter Wrens, especially males, who neither incubate nor brood, sleep singly in nests that they have built as well as in such miscellaneous cavities as pockets in thatched roofs and haystacks, clefts in old walls, unused nest boxes, chinks between thick ivy stems, old woodpecker holes, unfinished nests of Long-tailed Tits, and even old coconut shells. As the temperature drops with the approach of winter, these wrens abandon their solitary lodgings to clump together in sheltered places—a development unknown in tropical wrens, to which we shall return in Chapter 11.

Northern House-Wrens breed from southern Canada over much of the United States and in Northern Baja California in Mexico. Less hardy than Winter Wrens, which apparently have been for a longer age established in high northern latitudes, these wrens withdraw in autumn from all the more northerly parts of their breeding range, and their migratory habit makes family bonds less enduring than in many other wrens. After fledglings leave the nest box or other cavity in which they were reared, their parents lead them, as day ends, to some sheltered spot, which may be another nest box, the unoccupied nest of an American Robin or some other open cup, or the dense foliage of evergreen trees or shrubs. Rarely do they return to the nest they have just left. However, at dusk of the day a brood abandoned a bird box in Maine, their mother was heard chattering and calling near the box. Then she flew repeatedly between her fledglings and the box, guiding them to it. After all had entered, she joined them for the night. On the following evening, she had more difficulty gathering her little flock together. On the third evening, she tried without success to assemble her scattered brood. Other broods of Northern House-Wrens may sleep together for a week or more; but parents rarely feed, or remain associated with, their young for more than twelve or thirteen days after they quit the nest. Conducted by their elders beyond the parental territory, they soon disperse.

Widespread over the United States, except the eastern seaboard and the northern states, and in the highlands of Mexico, Bewick's Wren is resident over most of its range, including southwestern British Columbia. With a prominent white eye-stripe, it resembles the Carolina Wren, but is distinguished by its whitish underparts and the white outer feathers of its long tail. In habits, however, it is much more similar to the Northern House-Wren, with which it competes for nest sites. Almost any hole or cranny, in a building, a tree, a stone wall, or the ground, as well as any empty box or vessel of suitable size, is chosen for its usually cupped nest, which both sexes build. As in other wrens, the female alone incubates her large sets of four to eleven, most often seven, white eggs prettily spotted with shades of brown and lavender, which hatch in about fourteen days. For another two weeks, the young remain in the nest, fed by both parents. Fledglings are said to return every night to sleep in the nest, but details are unavailable. In California, an adult slept throughout the winter in a nest box provided for it. In the same state, Bewick's Wrens passed autumn and winter nights in crevices between thick slabs of redwood bark applied vertically to the sides of a house, perching on a wire against the wall of the dwelling, and in pockets amid thick accumulations of needles on a low bough of a Monterey Pine.

In humid lowland and foothill forests from southern Mexico to Peru and the Guianas lives a small, stubby-tailed wren whose loud, clear, melodious whistles suggest a larger bird. In pairs at all seasons, often with well-grown young, the White-breasted, or Lowland, Wood-Wren forages for small invertebrates in the deep shade near or on the ground. It builds two kinds of nests. Those intended for breeding are ovoid structures higher than wide, with firm, thick walls composed of leaf skeletons, rootlets, mosses, liverworts, and other vegetable materials. The round, sideward-facing entrance is shielded by a visorlike projection from the roof. Situated from ground-level to about 2 feet (60 centimeters) up, amid lush, screening foliage or in tangles of fallen branches and vines, these nests with two white eggs are rarely found.

The White-breasted Wood-Wren's second kind of nest is probably the most frequently noticed of all birds' nests in the lower levels of the forest, for it is often about head-height, rarely as high as 10 feet (3 meters) or as low as 2 feet (60 centimeters), and situated with little attempt to hide it in the fork of a slender sapling or in a tangled skein of vines. These nests are thin-walled pockets with the long axis horizontal and the opening at one end. Lacking a sill, they could hardly hold an egg. At night, my flashlight has often revealed a white breast gleaming in the doorway, or, if the sleeper awakes in the beam, a boldly-patterned black-and-white face looking outward. I have never found more than one adult in any of these dormitories; but in June of a year when the wood-wrens appeared to have enjoyed a particularly successful breeding season, I discovered two families in which two juveniles, with yellow corners on their mouths, lodged with an adult, doubtless their mother. About 50 feet (15 meters) from one of these nests with three sleepers was a similar nest with a single occupant, probably the male parent. After July, I no longer found more than one White-breasted Wood-Wren in a dormitory.

As one ascends from the lower slopes to the humid, moss-draped upper levels of montane forests over much of tropical America, the White-breasted Wood-Wren is replaced by a very similar bird with a gray breast. Gray-breasted, or Highland, Wood-Wrens live throughout the year in pairs, whose members sing delightfully to each other while they search for small creatures amid dense undergrowth. These wood-wrens build, often amid plants overhanging a roadway, path, or ravine, only one type of nest, a roughly globular structure with thin roof and walls composed largely of fine, dark rootlets and green moss. The roof that covers the rounded chamber extends far forward and downward, enclosing an antechamber or vestibule. The entrance opens downward, so that the wren must fly sharply upward to gain the interior of its well-enclosed nest. Despite the thinness of the roof, it sheds water well. In the cool, damp mountain forests, up to at least 10,000 feet (3,050 meters), mates sleep together at all seasons, except when breeding. If the nest where they have been lodging receives the usual two white eggs, the male often moves elsewhere; but in one nest he slept with his partner through most of the incubation period and all of the nestling period. After a successful nesting, fledglings sleep with both parents, making four occupants of a nest, which may be other than that where they were reared. Soon after the juveniles become independent, they appear to mate and build nests for themselves.

On the Pacific slope of southern Costa Rica and adjacent Panama, from lowlands up to about 4,000 feet (1,200 meters) lives a wren with bright chestnut upperparts, a long white superciliary stripe, and underparts everywhere finely barred with black and white. Throughout the year, family groups of three or four Riverside Wrens forage for insects in thickets along the banks of streams or at the forest's edge, proclaiming their presence by a large repertoire of songs that ring out above the clamor of a mountain torrent. Above a stream or its shore, or sometimes in a tree or shrub in a neighboring clearing, these strikingly patterned wrens saddle their nests upon a horizontal twig at no great height. On one side of the support is an ample, well-enclosed chamber; on the other side, balancing the chamber, is a vestibule or antechamber with a wide entrance that faces downward and often also inward. Whether used for breeding or sleeping, the nests are similar in shape, but the latter are often less carefully made, with thin roof and walls of fibers and green moss. While incubating her two white, faintly speckled eggs, the female sleeps alone, her mate in a dormitory nest. One evening I watched three Riverside Wrens, evidently a parent with two well-grown young, fly downstream, singing brightly, and enter a nest 15 feet (4.6 meters) above the water. More often a dormitory is occupied by a single wren, whose barred breast feathers, widely spread, fill the entrance to the inner chamber. Riverside Wrens build and lodge in their nests throughout the year.

The Carolina Wren, ranging from southeastern Canada over eastern and east-central United States into Mexico and northern Central America, is the only species of a large tropical genus, *Thryothorus,* that extends far into the North Temperate Zone. It exhibits a fascinating combination of traits inherited from tropical ancestors and adaptations to a region where winters are often harsh. Like its tropical relatives, it lives in pairs throughout the year and is not migratory. It pays for its sedentary habit by a drastic reduction of its population in colder winters when for weeks deep snow covers the ground where it chiefly forages. Instead of building its nests in trees and shrubs, like the Riverside Wren, Rufous-breasted Wren, and Plain Wren, it often places its bulky covered structures with a side entrance in a hollow in a tree or stump, a hole in a bank, amid the upturned roots of a fallen tree, or on the ground beneath exposed roots. These, its original sites, are still favored in wild woodlands; but now the Carolina Wren frequently chooses bird boxes, baskets, pails, discarded cooking vessels, or almost any receptacle of suitable size, as well as nooks and crannies in or under houses and outbuildings. Instead of the two or three eggs of *Thryothorus* wrens nearer the equator, it lays five or six, much more heavily marked with shades of reddish brown on a white ground than those of its tropical relatives. It needs to raise larger families to compensate for its higher mortality.

The male Carolina Wren helps to build the nest, feeds his incubating mate, and joins her in nourishing the young. While one female incubated in a basket on a porch in Arkansas, her mate slept in a fold of an awning outside a window. After the breeding season, these wrens apparently never build dormitory nests in the manner of the tropical species but seek relief from winter's cold in the most diverse situations, including, apparently not infrequently, old hornets' nests. A pair enlarged the opening in the side of a very big nest of White-faced Hornets that had been hung up in an outbuilding. During several winters they slept in it, and one spring they built a nest of their own in the top of this vespiary. Another pair slept with their tails protruding from a fold of an old portiere hanging in a garage. A pair lodged together in an old breeding nest. A single Carolina Wren snuggled on chilly September nights into the pocket of an old coat hanging on a porch. Another rested between two timbers on the inside of a garage wall. One spent the night in an old wren's nest in a box near the ceiling of a porch. Still another sought shelter on a March night in a pile of old cedar boughs, although buildings where it might have lodged were near.

From autumn until late in the following March, an enterprising Carolina Wren slept in a conservatory in New Jersey, at first entering and leaving through an open door, later through a hole made for his convenience. All winter he slept, ate, drank, bathed, and sang in the warm conservatory, from which he sometimes entered the dining room to gather crumbs. I have been unable to learn whether fledglings are put to bed in a nest or some substitute for one. On the probability that parent Carolina Wrens, like those of many other species in this family, install them in a suitable shelter, I include these wrens in the present chapter. Their miscellaneous lodgings serve the same ends as the nests built by their cousins in milder climates, justifying their inclusion among birds that sleep in dormitories at all seasons.

In rain forests from northeastern Honduras to western Ecuador, one sometimes finds a number of curious nests, shaped like bent tubes, resting conspicuously in upright forks of neighboring slender saplings. On one side of the crotch is a well-enclosed chamber; on the other, balancing the first, a downwardly inclined vestibule or entrance-way. Composed of lengths of vines, fibrous roots, and leaf skeletons, some of these bulky, elbow-shaped structures are in good repair, others in various stages of decay. As daylight fades beneath the massed foliage of tall trees, four or five grown Song Wrens, foraging in the ground litter to the accompaniment of loud, clear, well-spaced whistles mixed incongruously with guttural croaks, approach and fly up, one by one, to sleep through the night in a nest without eggs or nestlings. Song Wrens' breeding nests are of similar construction and usually contain two white eggs. The tenants of a dormitory nest are apparently parents with self-supporting young, but much remains to be learned about the habits of these shy forest-dwellers.

A very different dormitory-user is the Marsh Wren (also called the Long-billed Marsh-Wren), widely but unevenly distributed over much of the United States and southern Canada. Amid tall cattails, bulrushes, or reeds of freshwater and brackish marshes, a male builds in a single summer from five to thirty-two oval nests, each higher than broad, with a round doorway in the side. After completing the shells of several nests, a male, singing profusely and displaying with his cocked tail almost touching his nape, tries to entice a female to them. If she approves of the songster and one of his proffered structures, she may line it with softer materials, or she may build another nest in his territory, with little or much help from him. She alone incubates three to ten, most often five or six, dull brownish eggs, spotted with darker shades of brown, while he tries to lure another female to another of his constructions. The number of males that achieve polygamy varies with locality from 2 to 50 percent, but few have more than two females simultaneously. In the state of Washington, Jared Verner found males bringing a substantial fraction of the nestlings' food, whereas in Michigan Wilfred Welter never saw one feed a nestling.

After Marsh Wrens leave the nest at ages of fourteen to sixteen days, the brood may remain intact while fed by both parents, or it may be divided, each parent taking charge of one or more fledglings. Frequently the young wrens return to sleep in the nest where they were reared, or they lodge together in one of the unlined nests that their father built for courtship. Parental care continues for about two weeks after the young fledge, after which they join other broods that wander through the marsh. Adults also sleep in nests not used for breeding (whether sometimes a parent with its fledglings, I do not know); and in northwestern Washington, where in a maritime climate Marsh Wrens reside permanently, they lodge in these nests throughout the winter. Migratory populations of the same species, such as those in Michigan, appear not to use their nests as dormitories.

Herbert W. Kale II hand-reared nestling Marsh Wrens from the salt marshes of Georgia. When one of his fosterlings was a fledgling about a month old and still being fed by hand, it begged food from Kale and delivered it to two younger fledglings, thereby becoming a juvenile helper. In the evening, when the room where the wrens were kept was gradually darkened, the young helper flew to a nest of a Seaside Sparrow in a clump of grass, a favorite resting place. Settling into the nest, the helper began a subdued, rapid twittering, which had not previously been heard from a Marsh Wren. One by one, four younger wrens in the room approached the nest and snuggled into it with the helper, who did not cease to twitter until the last of them had joined him. Probably by similar calls parent Marsh Wrens draw their young brood to the nest chosen for their repose, but whether they do so in the wild marshes has not, to my knowledge, been disclosed.

In the earlier literature, the eggless, frequently unfinished nests built in numbers by Marsh Wrens and some others were often called "dummy nests," presumably made to mislead predators, which while searching in them for eggs or young would miss occupied nests. After it was learned that many wrens, especially in the tropics, build dormitory nests that never receive eggs, the dummy-nest theory lost credibility. When it was discovered that polygamous Marsh Wrens, Winter Wrens, and others build many nests to attract females, the notion that these structures were intended to deceive predators was further weakened. However, there is no good reason why the same nest cannot serve several ends. We have seen that eggless nests of at least certain populations of Marsh Wrens serve as dormitories for old and young. Moreover, we read in Bent's "Life Histories" of a Garter Snake that crept from one to another Marsh Wrens' nest while silent adults, with outspread wings and ruffled feathers, darted repeatedly at the serpent's head. After the snake had investigated several nests, it was killed and its stomach opened, revealing only a well-digested slug. The nests that it had entered were found to be unlined and empty. If the snake had not been destroyed, it might have abandoned an unrewarded search before it found a nest with eggs or nestlings. Although the multiple nests of certain northern wrens may serve a triple function—to attract mates, to deceive predators, and as dormitories—the habit of making them is probably derived from tropical ancestors who built them for dormitories. Male tropical wrens are not known to construct nests to attract mates (when already mated, they help their partners to build); nor, with one known exception, do these wrens build eggless nests in concentrations that would be likely to confuse predators searching for eggs and nestlings.

The arid lands of the southwestern United States and northwestern and central Mexico are the home of big Cactus Wrens, who build, preferably in cholla cacti, undeterred by a formidable array of sharp thorns, foot-long (30-centimeter) nests with bulbous chambers, each entered through a more or less elongated, sideward-directed spout. Where available in the desert, dry grasses are chosen for the nest's shell, which is lined with feathers. Near human habitations, scraps of paper, rags, string, cotton, and, above all, chicken feathers are incorporated in the structure. Throughout the year, adults sleep singly in these nests, those of paired birds usually not far apart. Where nests are frequently destroyed by Curve-billed Thrashers, the wrens, to avoid becoming homeless, continue to build throughout the winter. A dormitory nest may be used for breeding, or vice versa; but often new nests for eggs are built by both sexes working together. A female often sleeps in her breeding nest for seven to eleven nights before she lays the first of her three to five eggs. She alone incubates for a period of sixteen days, during which her mate may feed her three or four times daily. Nourished by both parents, the young remain in the nest for from nineteen to twenty-three days.

At the end of the fledgling Cactus Wrens' first day in the open, the parents, singing freely and flying ahead to show the way, lead them to a bedroom, often the nest in which they were hatched. If a fledgling, still with wavering flight, misses the narrow opening and alights on top of the nest, it delays there, appearing afraid to descend to the doorway. Then the parents climb up to the youngster and down into the nest, again and again, until, with this encouragement, the timid fledgling follows, sometimes tumbling onto an adult's back for help in getting in. For two or three weeks after the young wrens fledge, their parents continue to guide and help them to enter a nest, which need not be the same every night. Sometimes the mother or the father relinquishes her or his dormitory to the young wrens and sleeps elsewhere—apparently, parents and fledged young rarely, if ever, sleep together. At ages of fifty-two to seventy days, antagonism between brood members at bedtime causes them to lodge apart. When about three months old, they start to build dormitory nests for themselves. Rarely, juveniles of the first brood permit fledglings of the following brood to share a nest with them. Occasionally, Cactus Wrens little more than two months old give a few

Cactus Wren nest with fledglings about to enter

morsels to fledglings of a later brood—an inkling of behavior fully developed in some of the Cactus Wren's tropical cousins.

A neighbor of the Cactus Wren in the arid deserts of the southwestern United States and northern Mexico, the Verdin is a tiny, insectivorous, gray bird, with a yellow head and inconspicuous red shoulder patches. Formerly classified with the titmice in the family Paridae, it is now placed with the penduline tits of the Old World in the Remizidae—the only representative of this family in the New World. Its nests are more similar to those of certain members of the Neotropical ovenbird family than to those of any other bird of temperate North America.

Built by both sexes in a cactus or a usually thorny bush or tree, at no great height, the Verdins' nests are of three types: breeding nests, large dormitories, and small dormitories. The breeding nests, globular or ovoid structures about 8 inches (20 centimeters) high and broad, are composed of tightly interlaced, thorny twigs, so arranged that their ends project in a bristling array that repels predators from the small round doorway that faces outward from the supporting plant. One of the larger nests may contain over two thousand of these twigs, each from 2 to 5 inches (5 to 13 centimeters) long. Cobweb and woolly fibers fill the interstices among the twigs and bind them more firmly together. The chamber is well lined with spider web, grass blades, and pieces of dead leaves, within which is a layer of vegetable down and silken cocoons. Finally, the room is almost filled with small downy feathers, upon which rest three to six, most often four, bluish green eggs, finely spotted with reddish brown. In spring, when the desert is still rather cool, the nests face away from the prevailing winds; but in the hot summers they open toward the breeze, which cools the interior.

Large dormitory nests resemble breeding nests but have a smaller, shallower chamber that is often devoid of lining. They are occupied chiefly on winter nights; but if they remain sound enough, the male may lodge in them during the following breeding season. Small dormitories, built mainly in summer, are thin-walled, unlined or sparingly lined structures, sometimes open at both ends for the readier escape of a threatened sleeper. Adults appear always to sleep singly. Fledglings return to rest in their natal chamber for a number of nights before they build small dormitories for themselves.

Although adult titmice often sleep in holes in trees or other enclosed spaces, especially on cold winter nights, I have found no record of their leading their fledglings to a dormitory, with one exception, the Highland White-bellied Pied Tit of tropical East Africa. Adults sleep in knotholes and other cavities in trees, which are sometimes shallow but usually a few inches deep, in crevices between two stout upright trunks, or under loose sheets of bark, all usually high above the ground but occasionally as low as 4 feet (1.2 meters). They breed in deeper holes, sometimes old nests of woodpeckers or barbets. Four eggs were laid in a cavity in the gnarled and twisted trunk of an old olive tree in Kenya. In the evening after the young tits emerged, V. G. L. Van Someren watched the parents coax them back into their natal crevice for the night. Half an hour later, their father squeezed in with them; apparently, the mother had already entered. How long juveniles of the Highland White-bellied Pied Tit continue to lodge with their parents is not known. In England, parent Great Tits guide their fledglings to roost amid foliage in a tree, which seems to be the usual procedure in the titmouse family.

In South Africa, Red-winged Starlings build mud-walled cups in fissures in cliffs, clefts in boulders, holes in trees, ledges or nooks in buildings, or on shelves specially provided for them in porches. Pairs often sleep throughout the year in the sheltered nest site. After leaving the nest at ages of twenty-two to twenty-eight days, the two to four fledglings of the first brood return to sleep with their parents until five to seven days before their mother starts to lay a second set of eggs, when they are driven away. Like Southern House-Wrens and other young birds, they resist eviction from the familiar safety of their childhood abode, and for several successive nights they stubbornly persist in entering, sometimes provoking their father to resort to violence. Juveniles of the second brood are permitted to lodge longer with their parents, often for five or six weeks, before they are similarly expelled.

From Panama to Peru and Guyana lives a nocturnal bird known as the Oilbird because its nestlings, becoming extremely fat, have been slaughtered in great numbers as a source of cooking oil. In Spanish-speaking countries, it is called El Guácharo. About 18 inches (46 centimeters) long, with a wingspan of 3 to 3.5 feet (91 to 107 centimeters), it is largely bright brown, with white, black-margined spots on the crown, wing coverts, and outer tail feathers, and narrow, dusky bars on much of its plumage. Its large eyes are red, and its short bill is strongly hooked. In small or very large colonies, Oilbirds inhabit caves and large clefts among rocks, from seaside cliffs to high in the Andes, where they nest and pass their days, to sally forth at night and pluck on the wing, from palms and dicotyledonous trees, the large, often oily fruits that sustain them.

The cave where, unexpectedly, I made the acquaintance of these strange birds, of which I had read long before in the books of Alexander von Humboldt and Charles Kingsley, was situated among abrupt limestone foothills of the Peruvian Andes, through which the Río Huallaga and its tributary, the Río Monzón, carry their muddy currents down to the great Amazonian plain. The cavern, I was told in the settlement of Tingo María, was inhabited by flocks of "owls"; but I suspected from the verbal accounts that these cave-dwelling, gregarious "owls" were Oilbirds.

The hurried itinerary of the government mission with which I traveled allowed time for only a brief visit to the cave. A long afternoon walk through tall forests, freshly made clearings nearly impassable with fallen trunks, and abandoned patches of cultivation lushly overgrown with huge-leaved heliconias and shellflowers, brought us to the foot of a scarped mountainside, verdant with luxuriant tropical vegetation. By a short but difficult scramble over sharp-edged limestone, we reached the wide mouth of the cavern. Entering, we severed contact with the world of common, daily experience and passed into a realm so weird, so unearthly that we seemed to have been transported to another plane of existence. Neither Humboldt nor Kingsley, excellent though their descriptions are, had quite prepared me for the wild strangeness of a Guácharos' cavern.

Walking slowly inward between columnar stalagmites, from

vaulted chamber to vaulted chamber, penetrating into regions of dimmer and still dimmer light, we entered the dark inner sanctuary where the Guácharos rested in greatest numbers. The ground beneath our feet was soft and yielding. Casting downward the flashlight's beam, we found it carpeted with an undetermined thickness of large seeds regurgitated by the birds, the topmost round and fresh, those below ever farther advanced in decay. Many had germinated in the moisture of the cavern's floor, and in that nocturnal light had sent up crowded shoots, knee-high and of ghostly slenderness and pallor—a charnel house of sprouting life. Strange insects, vaguely glimpsed, scuttled out of the beam of light.

The floor of the cavern was the fit counterpart of its roof. The torchlight, now directed upward, revealed that the high, stalactite-studded ceiling and all the surrounding walls had weathered irregularly, leaving many projecting shelves and deep recesses in the rock. These standing-places were crowded with a weird assemblage. The beam of my small flashlight picked out, at that great distance, scarcely more than the ruby eyes of dusky birds dimly seen: innumerable pairs of gleaming red eyes, crowded companies of eyes, long ranks of eyes, shining from all the high, inaccessible ledges. So eery was the effect of all those disembodied eyes, staring down at me from far above, that had I ventured into that cavern alone, with no memory of the accounts of the naturalists who had made similar explorations before me, I might have been alarmed. It was easy to understand the superstitious awe in which the Indians of Caripe held the vast cavern of Guácharos visited by Humboldt and his companion, Aimé Bonpland.

At times the eyes, taking fright at the sudden illumination, left their places on the ledges and wheeled around in the darkness above us. Beneath that high, spreading ceiling was ample space for many birds to fly. As ledge after ledge was cleared of eyes under the disquieting gaze of my electric torch, the space above us filled with wide-winged, circling birds. Their harsh, raucous cries filled all the chamber with deafening uproar. But ever the beam, seeking out fresh recesses in the high walls, revealed more and yet more ruby eyes, gazing upon that wing-filled pandemonium of noise. When they swooped lowest in the flashlight's beam, these harsh-voiced fowls looked very much like overgrown nighthawks, Whip-poor-wills, and other birds of night. Years later, Donald R. Griffin proved that Oilbirds, like bats and some cave swiftlets of the genus *Collocalia,* can guide themselves by echolocation in utter darkness; but could they do so amid so much disturbing sound from a multitude of other birds?

When at length, tearing ourselves away from the unforgettable scene, we reached the mouth of the cave, the Sun had fallen behind the rounded limestone crests beyond the Río Monzón. We looked over the feathery crowns of palm trees upon soft clouds rose-tinted in the sunset. How quiet and restful seemed the forest, after the factorylike din we had left behind us! But doubtless the inner chambers of the cavern had returned to silence and repose, now that the disturbing flashlight had been removed. Already, green parakeets with red foreheads had gathered in the trees around the cavern's doorway, evidently only awaiting our departure before they settled down in its shelter for the night. Soon the Guácharos would be streaming forth in their thousands to seek their breakfast of tree seeds. A canoe had come up the Monzón to meet us; and as we glided down the silvered water in the moonlight, we watched the dusky birds winging in silence over the treetops.

Oilbirds are included in this chapter because, as David W. Snow learned in Trinidad, throughout the year, whether breeding or not, they spend their days in pairs on their nests, built upon ledges and in recesses in dimly lighted or wholly dark caverns. Resting side by side, facing outward, with bodies tilted slightly forward, they sleep at intervals, and when they wake the male sometimes preens his partner's head. The nest is a broad, shallow saucer, about 15 inches (38 centimeters) in diameter, rimmed with regurgitated fruit pulp. On it the female lays two to four white eggs, which she and her mate, sitting alternately, incubate for 33 or 34 days. Hatched almost naked, the nestlings, like those of certain swifts, later sprout a downy coat. Nourished with regurgitated whole seeds, they put on so much fat that, when about 70 days old, they weigh 50 percent more than adults. This excess is lost before, at the age of 90 to 125 days, they fly from their ledges. Sometimes they return to rest upon their nests by day, but how long their association with them continues is not known.

With few exceptions, parents of nidicolous birds continue to feed them for days, weeks, or months after they leave the nest. Only a minority put their fledglings to bed in dormitories where they sleep more warmly and safely; even some that themselves lodge in nests or other closed spaces leave their young to pass the night in the open. Tucking their fledglings into bed is the last refinement of parental care in birds, practiced only by parents most solicitous of their progeny's welfare. That the juveniles are sensible of the advantages of sleeping under cover, in contact with or near their parents, is evident from their stubborn resistance to expulsion, when they are well able to take care of themselves and it is time for them to depart. Some, indeed, remain with their parents for one or several years, with interesting consequences to be examined in the following chapter.

References: Anderson and Anderson 1957; 1962; Armstrong 1955; Austin 1976; Bent 1946; 1948; Griffin 1974; Howard 1952; Kale 1962; Kendeigh 1941; Miller 1941; Nice and Thomas 1948; Preble 1961; Rowan 1955; Skutch 1940; 1953; 1960; 1981; 1983b; Snow 1961; Tennessee Ornithological Society 1943; Van Someren 1956; Verner 1965; Welter 1935; L. Williams 1941.

TEN

The Dormitory Becomes the Family Dwelling

Beginning with female birds who opportunistically sleep in their breeding nests before they lay eggs in them and after their young have flown, but build none solely for lodging, we have traced the development of the dormitory habit to successively higher levels. An advance is made when both parents sleep in the breeding nest, and a further advance when fledglings are led back to it. When, in addition to breeding nests, birds build or otherwise acquire nests or enclosed spaces similar to those in which they rear their young and occupy these bedrooms throughout the year, they take a long step forward. They may sleep in these dormitories alone or in pairs. When parents are careful to provide for their fledglings the same nocturnal protection that they themselves enjoy, the dormitory habit attains an advanced level. When, finally, instead of spontaneously or compulsorily separating from their parents soon after they can nourish themselves, the young continue to lodge with their parents long after they can support themselves, often for one or several years, the dormitory habit rises to its apogee. At this seventh stage, the dormitory may, without exaggeration, be called the family home.

Many of the species that have attained this highest level of dormitory-using breed cooperatively, or communally, as this breeding system is also called. Cooperative breeding is distinguished by the presence of helpers, most often nonbreeding birds who assist a breeding pair in defending their territory, feeding and protecting their nestlings and fledglings, and, often, helping to guide the latter to their bedroom. Less frequently, the auxiliaries participate in nest construction, incubate eggs, or brood nestlings. The helpers are more often males than females, who among birds have a stronger tendency to disperse at an early age. Usually they are yearlings or older offspring of the pair they aid, who contribute to the support of their younger siblings; but not infrequently the cooperating group is joined by individuals from other families. In typical cooperative groups, the breeding male is head of the family, with other members ranking below him in descending order of their ages. However, the means by which this hierarchy is maintained are restrained rather than harsh interactions between group members, who dwell together in harmony. Cooperative breeding in birds is one of the highest manifestations of social life among vertebrate animals, equaled by only a few mammals, and hardly exceeded by humans except in the most closely integrated families.

In Chapter 4 we noticed that cooperative breeders often roost in closest contact on a branch. In the present chapter we meet cooperative breeders who build covered nests or occupy holes or burrows and sleep with equal or greater intimacy in these enclosures. Whether such birds lodge together because they cooperate, or cooperate because they lodge together, I cannot say; but undoubtedly these two behaviors are closely related. Because the perpetuation of most species of birds depends upon direct feeding of the young, to place food into gaping avian mouths has become one of the strongest, most persistent of avian impulses, often arising in immature individuals, and not wholly extinguished even in some nest parasites that never raise their own young, such as certain cuckoos and cowbirds. Accordingly, it is not surprising that birds who continue to dwell with their parents while the latter rear subsequent broods respond appropriately to the begging of younger siblings with whom they are so closely associated, thereby becoming helpers at the nest. Cooperative breeding and sleeping together are two aspects of the same closely integrated pattern of living.

While surveying successive stages in the development of the dormitory habit, we have jumped from family to family instead of trying to trace its evolution in any single family. We do not know enough about the sleeping habits of birds to attempt to follow their evolution in a single lineage. Families that exhibit a number of levels of the dormitory habit and, accordingly, are promising candidates for the study of its evolution, are woodpeckers, wrens, and weavers, all of which we shall meet again in the following pages.

Before we return to these great families, let us look at the wood-hoopoes, a small family of nine species confined to forests and savannas of Africa south of Sahara. Ranging across the breadth of equatorial Africa and southward to the Cape of Good Hope, Green Wood-hoopoes are long-tailed, 13-inch (33-centimeter) birds, whose dark plumage, glossed with green and blue, is relieved by white patches on wings and tail. Their long, slender bills, short legs, and strong toes are red. They climb over trunks and branches, propping themselves with their long tails, while their pointed bills probe crevices in bark and wood for small invertebrates. Versatile foragers, they also search among foliage and twigs, and on the ground.

For three years, David and Sandra Ligon studied wood-hoopoes on the plateau of central Kenya, where the birds lived permanently in open, parklike groves of acacia trees, usually in cooperating groups of three to nine individuals, less often in single pairs or parties of up to seventeen. While the single breeding female of each group incubated her two to four plain white eggs for eighteen days, she was fed by from one to seven helpers as well as by her mate, the dominant member of the family. During the month that the young remained in the nest cavity, they were nourished chiefly by the helpers, who might bring up to 83 percent of their meals. At an exceptionally late nest in January, the nestlings were fed by nine adults plus three juveniles fledged in July, a total of twelve attendants.

Throughout the year, Green Wood-hoopoes lodge, singly or up to seven together, in cavities in trees, either old woodpeckers' holes or hollows resulting from decay. Apparently, they need these shelters to keep warm during the cool nights of the plateau, 6,400 feet (1,950 meters) above sea level. When disturbed during the night, wood-hoopoes turn their tails toward the doorway of their dormitory, and from their exposed oil glands expel drops of a fetid oil that may repel predators or perhaps the abundant Acacia Rats that covet their holes. Despite this defense, wood-hoopoes often lose their dormitories to the rats, to pugnacious Blue-eared Glossy Starlings, to aggressive African honeybees, and to other creatures. When this occurs, they sometimes sleep with no better shelter than loose sheets of bark.

After the young Green Wood-hoopoes fledge, all group members continue to feed and protect them. Late in the afternoon, the adults lure the fledglings to the dormitories selected for them by bunching closely together, swaying their bodies from side to side, and tossing their tails up and down in a ludicrous fashion, while they join their voices in a chorus of loud, excited, chattering notes. Attracted by this "rally," the youngsters approach their elders, who then fly ahead and repeat the display, calling the inexperienced birds onward by stages. When, with much excitement, the family reaches the chosen cavity, an adult enters and the young follow. Still hungry, they beg and are fed inside the hole. After an interval of foraging for themselves and quietly preening, some of the adults may join the fledglings for the night. For about two weeks after their emergence from

Green Wood-hoopoe at nest

the nest, the young wood-hoopoes are put to bed in this manner, not always in the same cavity. Often they alternate between two or three holes, probably for greater safety from predators.

Another African bird that breeds cooperatively is the White-fronted Bee-eater, which in Lake Nakuru National Park in Kenya Stephen Emlen and his associates studied for three years. Here these colonial birds live in extended families or clans that comprise up to sixteen individuals of three or occasionally four overlapping generations. Throughout their lives, clan members preserve close bonds. In densely crowded earthen cliffs, they cooperate in digging burrows 3 to 6 feet (0.9 to 1.8 meters) long, in which they nest and sleep together throughout the year. After flying from their burrows on cool mornings, members of a single family often huddle together on a branch. From one to five helpers may assist a breeding pair, feeding the female, incubating, nourishing and protecting the young.

In the rainy mountains of Costa Rica and western Panama lives a small, stout bird, plainly attired in golden-brown, olive, and gray. Its short, thick bill bears at the tip of the upper mandible a fine prong that fits between two similar projections on the lower mandible. Prong-billed Barbets wander through mossy forests and adjacent clearings with scattered trees in families or small groups that consume great quantities of fruits and flowers, supplemented by a few insects, and join their voices in loud duets and choruses. With bills that seem ill suited for such work, pairs carve deep, woodpeckerlike cavities in fairly sound dead trunks. While incubating, for thirteen days, four or five plain white eggs, and while rearing the nestlings, both parents sleep in the nest cavity. By assiduously removing the young birds' droppings mixed with wood particles scraped from the walls for this purpose, they keep their chamber scrupulously clean. After the fledgling barbets' first flight, they are led by their parents to lodge in the nest cavity with them. I watched a pair trying to guide two fledglings back to the hole they had left earlier that same day. After an hour of flying back and forth between their offspring and the doorway and going in and out of the cavity, they succeeded in leading one of the weakly flying young to the bedroom. Then, daylight fading, both parents entered with the single fledgling, leaving the more backward sibling to pass a wet night in the open. On the following evening, both young barbets entered to sleep with their parents.

After juveniles support themselves, Prong-billed Barbets sleep gregariously, sometimes in holes carved by woodpeckers. Between August and March, I found up to sixteen of them, obviously not all members of a single family, lodging together. The hole in which the greatest number slept was so small that the birds must have rested in two or three layers. As the breeding season approaches, barbets lose their sociability, separate into territorial pairs that vigorously repulse intruders, and carve new nest holes, in which a male and female sleep as soon as the chambers are large enough.

In Africa, where barbets are much more abundant than in the New World, Carl Vernon learned that a group of Black-collared Barbets defends a territory throughout the year, and when not breeding all its members sleep together in the same hole. To reproduce, a group of eight split into subgroups of three, three, and two, each with its own nest; but the trios and the pair visited each other's nest trees. In Kenya, V. G. L. Van Someren watched four Pied Barbets that looked alike feed an undetermined number of nestlings. He found six grown birds sleeping together in a tree hole. In the Frankfort Zoo in Germany, nestling Double-toothed Barbets were fed by a second female and juveniles of the preceding brood. After the young fledged, the whole group continued to attend them, and they were led back to sleep in the nest cavity. In various parts of Africa, from three to six White-eared Barbets have incubated the eggs, brought food to a nest, and slept in the chamber with the nestlings. In South Africa, eleven birds of this species slept together in a cavity.

Confined to woodlands of tropical America, toucans attract attention by their enormous and often highly colored beaks. Largely frugivorous, they vary their fare with insects and occasionally eggs or nestlings of other birds. Although, as related in Chapter 4, big *Ramphastos* toucans, such as the Red-billed Toucans in Guyana, roost in contact high in trees, middle-sized araçaris prefer to sleep in cavities, often five or six in the same old woodpeckers' hole. To fit into a narrow, crowded chamber, they fold up, turning their heads over their backs and bringing their long tails forward to cover them. In a hollow in a thick, horizontal branch, entered through a narrow aperture on the underside, six Collared Araçaris slept high above the ground in a Panamanian forest. When eggs were laid in this cavity, all but a single incubating parent found lodgings elsewhere. After the nestlings hatched, all but one of the original six adults returned. All five of the grown birds who slept with the nestlings brought food to them, continuing through the six weeks that they remained in the nest.

The first fledgling Collared Araçari to emerge from the nest found it difficult to return through a downward-facing orifice so narrow that the adults could barely squeeze through it. In the evening, while the grown araçaris were showing it the way in the usual manner of preceding it to the hole, a White Hawk swooped down and car-

Collared Araçari in nest entrance

ried off the piteously wailing youngster. Because of this tragedy, the attendants did not try to lead the second and third fledglings back to their natal chamber but doubtless found other lodgings for them, as araçaris often have alternative holes to use in such emergencies. The related Fiery-billed Araçari of the Pacific slope of southern Costa Rica and adjacent Panama has similar sleeping habits and leads its fledglings back to the nest hole. It has recently been reported to have helpers.

In addition to the many woodpeckers that always lodge singly and were considered in Chapter 7, there are a few more social species. Among them is the Golden-naped Woodpecker, which inhabits the shrinking rain forests of the southern Pacific sector of Costa Rica and adjoining parts of Panama. This small black-and-white woodpecker has a red patch on the abdomen. The yellow forehead and nape are separated by the red crown of the male, the black crown of the female. Throughout the year, Golden-napes live in pairs and family groups. As soon as a male and female, working by turns, have finished carving a hole, high in a dead or dying tree in the forest or lower in a nearby clearing, they sleep together in it, continuing this practice during the twelve days that, sitting alternately, they need to hatch their three or four glossy white eggs, and the thirty-three to thirty-seven days that the young remain in the nest. Toward the end of the fledglings' first day in the open, the parents lead them back to the nest. Usually the young woodpeckers enter without much instruction; sometimes one who has been climbing about on the nest trunk and/or neighboring branches finds its way home without parental guidance. After the fledglings' return to the nest in the afternoon, the parents feed them there. The sight of siblings being fed often brings a laggard youngster promptly to join them inside. Later, both parents enter to sleep with their fledglings.

I have known Golden-naped Woodpeckers to attempt a second brood only when their first brood comprised only female fledglings, who continued to lodge with their parents while the latter incubated the next set of eggs and attended the nestlings. On one occasion, three juvenile females, thus closely associated with their younger siblings, brought a little food and tried incompetently to deliver it. After the later brood fledged, this hole held seven sleepers, including the two parents, three juvenile females of the first brood, and one male and one female fledgling of the second brood. Two weeks after the latter began to fly, their dormitory in a decaying branch fell. All slept in the long-neglected hole where the first brood had been reared, while they carved a new chamber for themselves. In this task the parents were assisted not only by the females of the first brood but also by the male of the second brood, who had graduated from the nest only three weeks earlier and was now about two months old. One morning, after all the rest of the family had suspended toil and departed, he remained alone, throwing out chips which he himself had probably loosened inside the cavity, where I could not see him.

If they survive, young Golden-napes frequently continue to lodge with their parents until, at the approach of the next breeding season in March, the mated pair move into the hole that they have just completed, where the adult female will soon lay. Then, when nearly a year old, the younger members of the family disperse, apparently without much urging by the parents.

In the eastern foothills of the Ecuadorian Andes and beside the upper Amazon, I have found up to five Yellow-tufted Woodpeckers, close relatives of the Golden-napes, sleeping in the same hole, which they entered after spectacular displays of flycatching in the evening air. These woodpeckers appear to breed in cooperative colonies, but their social system remains to be carefully studied.

A neighbor of the Golden-naped Woodpecker in southern Costa Rica is the Olivaceous Piculet, whose sleeping habits are quite similar. In a miniature nest cavity carved by both sexes in soft wood, sometimes a decaying fence post, these pygmy woodpeckers sleep in pairs while incubating their two or three tiny white eggs and rearing their nestlings. When, at the age of twenty-four or twenty-five days, the young leave the nest, they are guided back to sleep in it with both parents. As among Golden-napes, young of the first brood continue to sleep with parents rearing a second brood, but they have not been seen to feed their tiny siblings. The young may continue to lodge with their parents until at least three months old. Because falling or crumbling of the decaying stubs that contain piculets' holes makes them seek other shelters, I have not succeeded in following the fortunes of a family longer than this. In Ecuador, a male and two females of the Golden-fronted Piculet slept together in a chamber with eggs, apparently of a second brood. In Suriname, a young Arrowhead Piculet continued to lodge with its parents for at least two months after it fledged. Such prolonged family unity is probably widespread among the twenty-seven species of these diminutive woodpeckers in tropical America.

Still another woodpecker that makes of its hole a family dwelling is the Blood-colored, a pair of whom, watched by Fr. Haverschmidt in Suriname, lodged together for eight months, during part of which period a third individual accompanied them at night. In Thailand, Lester Short found a family of five Greater Golden-backed Woodpeckers who nightly slept, singly, in as many holes, situated one above another in the same trunk. This might be considered a family dwelling with individual bedrooms. Similarly, a male and two female Red-crowned Woodpeckers, apparently a pair with a grown daughter, lodged for months in separate holes close together in a slender stub. Although these woodpeckers insist upon sleeping singly, they are not averse to having family members lodge near them in separate bedrooms.

Dressed in a harlequin pattern of white, red, and glossy greenish black, with the female distinguished from the red-crowned male only by a black band across the top of her head, the Acorn Woodpecker is one of the most social species in the woodpecker family. From northwestern Oregon it spreads over far western and southwestern United States and the highlands of Middle America to western Panama, then, after a gap in its distribution, it reappears in the northern Andes of Colombia. Acorn Woodpeckers live in pairs or groups of up to fifteen individuals, the larger of which consist of breeding males and females and nonbreeding birds of various ages. Unless forced by scarcity to wander, the pairs or groups are highly sedentary on territories which all grown members defend. An adequate territory contains oak trees for acorns, granary trees for storing them, other trees for sweet sap, nest holes, and dormitory holes.

Olivaceous Piculets at nest hole

Although many other woodpeckers and birds of other kinds store food, these woodpeckers have a neater method. In the bole of a dead trunk or the thick bark of a living one, they carve holes, each just large enough to hold a single, snugly-fitting acorn. A large storage tree may be honeycombed with thousands of these little pits. Not all populations of Acorn Woodpeckers make these granaries so revealing of their presence. In humid tropical mountains, where fresh food is available throughout the year and stored mast would soon decay, some woodpeckers simply tuck whole or fragmented acorns into crevices in wood and bark or amid massed epiphytes.

Just as a dependable supply of food is the foundation of advanced human cultures, so the woodpeckers' granary of acorns is the indispensable support for their social system, at least in the north where winters are cold. Unless in autumn they lay up enough mast to last until the following spring, they may be unable to breed successfully. Each group or territory has at one time a single active nest, with usually three to five eggs, rarely as many as thirteen. The larger sets are produced by two females sharing a nest. In a territory with more than one male and female, the nest is usually attended by nonbreeding yearling and older helpers of both sexes, who participate in incubation, as well as feeding, brooding, and defending the young. When a number of birds share incubation, they may replace one another with surprising frequency; at a Costa Rican nest, the average length of 108 shifts on the eggs by five cooperating Acorn Woodpeckers was only 5.1 minutes. A single male took charge of the nest at night.

After leaving the nest, fledgling Acorn Woodpeckers are fed by most of the adults and yearlings of their group. I do not know whether they lead newly emerged young to a dormitory, as Golden-naped Woodpeckers and piculets do; probably, following their elders, the young soon find a suitable hole without much guidance. Up to twelve individuals, the entire membership of a group, have slept in the same chamber; but frequently a large group occupies several holes, a few in one and a few in another, with sometimes a solitary sleeper. On rainy days Acorn Woodpeckers spend much time in their dormitory holes, or they cling beneath branches for shelter. The approach of a hawk sometimes sends them into their dwelling, whose thick wooden walls protect them from the raptor. Their dormitory hole is their castle.

In the account of the sleeping habits of the Troupial in Chapter 7, we glanced at the nests of the Rufous-fronted, or Plain-fronted, Thornbirds, which this oriole pirates. About 6.5 inches (16.5 centimeters) long, dark brown above and whitish below, these plainly colored, wrenlike birds of the ovenbird family are distributed discontinuously in northern Venezuela, northern Peru, and from the hump of Brazil to Bolivia, northern Paraguay, and northern Argentina. Their nests are built of stiff twigs compactly interlaced by both members of a pair, sometimes with a little help from their older offspring. These bulky structures hanging from the ends of usually high branches nearly always consist of at least two chambers. By the addition of room above room, they may attain a height of about 7 feet (2 meters), with about eight or nine chambers, but such tall structures are exceptional. The rooms, about 5 inches (12.5 centimeters) wide and high, do not intercommunicate; each has its separate entrance from outside, usually through a narrow passageway or antechamber in front of the main chamber.

The floors of these rooms are lined with almost any soft or flexible materials that the thornbirds can find: strips of fibrous bark, fibrous pieces of decaying stems and leaf-sheaths of tall grasses, and, near human habitations, such miscellaneous trash as fragments of plastic bags, scraps of cellophane, colorful candy wrappers, tinfoil, bits of paper, and chicken feathers, as well as fibrous materials. In the lowest chamber, which is least accessible to a flightless predator advancing along the supporting branch, the female lays three plain white eggs, which she and her partner, sitting alternately by day, incubate for a period of sixteen or seventeen days. During the three weeks that the young remain in the nest, both parents bring them larval, pupal, and mature insects, rarely a spider, always carrying a single item in the end of the bill. From the beginning, both parents sleep in the nest.

Clad in plumage much like that of the adults, fledglings fly well when they leave the nest. In the evening, the parents lead them back to it by flying from them to the hanging structure and going in and out of one or more chambers. Young thornbirds do not need much urging to retire. At the end of their first day abroad, three fledglings followed a parent so closely that all four tried to enter together, jamming the narrow passageway. Then some came out while others tried to push in, causing much confusion. Finally, all settled down, to be joined later by the second parent. From the first, fledgling thornbirds spend a long day with their parents in the open, leaving with or soon after the adults in the morning, returning with them in the evening. Accordingly, unlike certain wrens and woodpeckers, they are not fed in the nest after their first departure.

Young thornbirds continue for months to sleep with their parents. In May and June, before any young-of-the-year had fledged, fourteen of twenty-two nests that I watched in Venezuela were occupied nightly by single pairs, who regularly passed the night together in the same room. Three of the nests had six occupants each, probably parents with the survivors of two broods of the preceding year. With so many sleepers, it was difficult to learn how they distributed themselves for the night, for some would continue to pass confusingly from room to room until the light had become so dim that I could hardly distinguish their dark forms as they crept rapidly over the outside of their dark nest from one entrance to another. Similarly, in the dim light before they flew down at dawn, they would often shift from chamber to chamber. At times one would force its way in with others who resisted its intrusion, when without opposition it might have entered another compartment of the same nest. Parents with eggs or nestlings sometimes tried to exclude older offspring from the brood chamber, not always successfully. Although it often took the thornbirds a while to settle down, after arranging themselves for the night they seemed to enjoy the intimacy of the nest. Standing beneath one of these structures at dawn, I would hear them twittering and chirping contentedly before they emerged.

In one nest that I watched, four nonbreeding birds of the preceding year slept while the parents raised the season's first brood. After this brood of two fledged and returned to this nest, it sheltered eight active thornbirds, the largest number that I found lodging in any nest. Soon after the first brood began to fly, two of the four nonbreeding grown birds who had been present at the beginning of the breeding season disappeared, and the number of occupants was again reduced to six. When they started to incubate their second set of eggs, the parents, becoming increasingly antagonistic to the other birds who continued to sleep in the nest, tried hard to exclude them in the evening. But the others, resisting expulsion from the home that they had occupied for so long, entered the upper compartment after their parents had retired into the lower one with the eggs. At break of the following day, the parents, in a milder mood, permitted the others to pass from chamber to chamber with no apparent opposition.

Rufous-fronted Thornbirds do not welcome guests in their dwellings with spare bedrooms; but individuals who have suddenly lost their home, as by the snapping off of an overladen supporting branch, are very persistent in demanding hospitality. The uninvited guests enter in the evening after the nests' builders have retired, leave earlier in the morning, and go their own way. One evening my wife and I watched a thornbird, whose nest had just fallen, try until well after nightfall to force its way into the nest of a neighboring pair. After repeated repulses, it stood in the doorway facing outward, as we could see by the light of the rising moon, probably receiving on its inwardly directed tail the pecks of the defenders. Finally, the intruder pushed inside until it was no longer visible.

The original occupants of a nest may become reconciled to the presence in an adjoining compartment of uninvited guests, but even after months of residence these birds do not become integrated into the family. Although the many-chambered nests of Rufous-fronted Thornbirds appear to be avian apartment houses, like the big nests of Monk Parakeets in Argentina and of Palmchats in Hispaniola, they are strictly family residences. I never found two pairs breeding in the same hanging structure.

The Firewood-gatherer, or Leñatero, is a larger member of the ovenbird family, about 8 inches (20 centimeters) long. Light brown above, it has a rufescent forehead, whitish eyebrows, white throat outlined by small black spots, and buffy breast and abdomen. In northeastern Argentina, Uruguay, Paraguay, and southern Brazil, it walks over the ground in open woodlands and weedy fields, gathering insects and other small invertebrates. The Leñatero receives its name from the number of sticks, often thorny, that it gathers for its nest low or high in a tree, for, in addition to all those that it incorporates into its massive structure, it drops many that it fails to retrieve from the ground beneath it, probably because it cannot fly straight up to the nest with them. Sometimes enough sticks to fill a wheelbarrow accumulate beneath a nest. The globular or cylindrical structure of tightly interlaced twigs may be used for at least two years and, exceptionally, become 43 inches long by 18 inches wide (110 by 45 centimeters). Entry is from the top, through an obliquely descending, curved or spiral tube, which in the larger nests may be 31 inches (80 centimeters) long. The rounded chamber in the lower part is thickly lined with feathers, seed down, wool, rabbit skin, snakeskin, and the like. In this well-enclosed room three to five white eggs are incubated by both sexes for sixteen or seventeen days, and the nestlings are fed for twenty-one or twenty-two days.

After young Leñateros fledge, they are led to sleep with their parents in the nest. According to W. H. Hudson, the juveniles continue to lodge in the parental home for three or four months; but Rosendo Fraga found that they were expelled from it less than one month after they fledged. The parents continue to sleep in it through the winter. Starting when about forty days old, the young Leñateros help their parents by adding sticks to the nest in which they still sleep. Builders of large and elaborate nests, like those of Rufous-fronted Thornbirds and Firewood-gatherers, continue to be occupied with their construction and maintenance as long as they are in use. In the rather similar nest of another Argentine ovenbird, the Crestudo, or Larklike Brushrunner, Fraga found eight birds, probably parents with juveniles, sleeping together.

In the western United States, where it is widely but patchily distributed, chiefly in the mountains, the bluish-gray-and-white Pygmy Nuthatch forages high in tall pine trees. In coastal central California, Robert Norris found three individuals in attendance at eight of thirty-six nests. The helper was always a nonbreeding male, usually a yearling but sometimes older, who had remained closely associated with his parents and could not find a mate because, for some obscure reason, males were much more numerous than females. The auxiliary helped to excavate the nest cavity in a tree and to line it with feathers, shreds of bark, rabbit fur, wool, plant down, snakeskin, and other soft materials. He fed the female parent during the sixteen days that she incubated her five to nine eggs. When they hatched, he helped the parents to feed the nestlings with insects and spiders, and he carried droppings from the hole. After the young emerged when about three weeks old, the helper continued to attend them until, at the age of about seven weeks, they could find enough food for themselves.

From the start of nesting, the male helper slept in the nest cavity with the parents and their eggs or young. At the end of the fledglings' first day outside, their attendants led them to the nest in which they were reared or to some other suitable cavity, sometimes as much as an hour and a half before the parents and helper joined them for the night. Next morning the young nuthatches might linger within for half an hour after the adults became active. During both of these intervals, the attendants brought food to them and removed their droppings from the hole. Other fledglings entered later in the evening, left earlier in the morning, and received no food while in their bedroom.

Pygmy Nuthatches raise only one brood in a year. After the breeding season, and throughout the winter months, parents, young, and the helper when present, continue to lodge together. In the colder months, members of other families are not denied admittance to the dormitory and may increase the number of sleeping nuthatches to eleven. Although in the relatively mild climate of coastal California families tend to remain intact through the winter, and Norris never found as many as a dozen sleeping together, in the much more severe winters at high altitudes in the mountains of the interior of the continent, many more Pygmy Nuthatches huddle together in a hollow trunk—a matter to which we shall return in the following chapter.

The big (8-inch or 20-centimeter) Banded-backed Wren ranges from eastern Mexico to western Panama and, after a gap in its distribution, appears again in northern Colombia and northwestern Ecuador. As versatile in choice of habitat as in modes of foraging, it lives in clearings with scattered trees as well as in woodland, in humid and arid regions, from lowland rain forests to montane woods of oaks and pines, and above them in cypress forests nearly 10,000 feet (3,050 meters) above sea level. It searches for insects and other small invertebrates in the ground litter, on the trunks of trees, and amid foliage on lofty boughs.

In Banded-backed Wrens the cooperative breeding foreshadowed by the occasional feeding of fledglings by juveniles of the related Cactus Wren is fully developed. These wrens live at all seasons in pairs or, more often, in groups of about six to twelve individuals, who throughout the year retire in the evening into large, roughly globular nests, often nearly a foot (30 centimeters) in diameter, with a wide side entrance protected by the overhanging roof. Composed of straws, pine needles, moss, lichens, sheep's wool, or whatever the wrens can find to make them dry and cozy, these nests are built by several members of a group; perhaps all adults help. They may be situated low in garden shrubs or 100 feet (30 meters) up in trees, often near the ends of slender branches. In wet lowland forests they are frequently hidden in the midst of luxuriant masses of epiphytes in treetops.

Three to five white eggs, immaculate or faintly speckled with brown, are laid in a nest which may be a weathered but still sound structure already used as a dormitory. As in other wrens, they are incubated by the female alone, while her mate and helpers lodge in another nest. Exceptionally, as in the case of a male who had lost his dormitory and was building another, he may pass a night or two with his incubating mate. The same sleeping arrangements are followed during the eighteen or nineteen days that the young remain in the nest, fed by helpers as well as parents.

After their departure, fledgling Banded-backed Wrens are led in the evening to the nest in which they were reared or to a neighboring one. I watched a helper show the young wrens how to enter a dormitory with a doorway difficult for them to reach because there was no perch in front. To rise to it from a twig directly below was almost beyond their strength, and when they alighted on the forwardly projecting roof, they saw nothing below to which they could conveniently drop. While their patient instructor went in and out, many times over, to show the fledglings the way, one of them sometimes clung to its back. The adult might have pulled the clinging youngster in, but this solution of the problem was neglected. With the helper's untiring guidance and encouragement, after many trials and failures the fledglings gained toeholds in the doorway and managed to clamber inside. The efforts of the little, short-tailed, pale-breasted, weakly flying fledglings to imitate their long-tailed, spotted-breasted teacher were endearing, and so amusing that, while watching them, I shook with silent laughter. After an interval, the three fledglings were joined by the three adults of this group. As each in turn darkened the entrance, the young wrens greeted it with lisping hunger calls, but they were not fed at this time.

At another nest in the Guatemalan highlands, four or more adults of a group of seven fed five nestlings. After these young fledged, they slept with the parents and helpers in a dormitory nest about 100 yards (90 meters) from the breeding nest. This made a family of eleven, the largest number of Banded-backed Wrens that I have found lodging together. From May until the year's end, this group (later reduced to nine) occupied only two dormitories, one of which was the brood nest. Between May and December, the smaller group of six wrens, far less settled, alternated between five different dormitories, some newly built, others old. The habit of sleeping together in snug dormitories helps Banded-backed Wrens to thrive from warm lowlands to frosty heights.

The related Rufous-naped Wren, an inhabitant of arid and semiarid country from central Mexico to northwestern Costa Rica, lives in family groups and apparently also breeds cooperatively, although no thorough study of its habits has been reported. These big wrens breed and sleep in bulky, pocketlike nests with a side entrance, often built in an opuntia or other cactus, or in a bull's-horn acacia tree swarming with fiercely stinging ants that repel creeping or climbing intruders. In Costa Rica, a family of four Rufous-naped Wrens entered farm buildings to hunt insects and spiders. With two nests in separate rooms of a dovecote on a post and another in a nearby orange tree, they were building a fourth, in November, far from their breeding season. If disturbed after they had entered one of their nests in the evening, they moved to another. A third species of *Campylorhynchus,* the Giant Wren, is confined to the state of Chiapas in southern Mexico, where in July I found their bulky nests prominent in bull's-horn acacias growing in bushy pastures. One had two doorways in the sides. While the female brooded two nestlings in this unusual structure, her mate slept nearby in a smaller, empty nest with a single entrance.

The Gray-crowned Babbler of eastern Australia is a 10-inch (25-centimeter), strong-legged bird with a fairly long, sharp, curved bill.

Grayish brown above, it has white eyebrows, dusky face, white throat and chest, pale brown abdomen, and a white-tipped tail. In open, grassy woodland, parties of these babblers forage for insects and other invertebrates on the ground, where they overturn small objects by inserting the bill beneath them and pushing forward and upward. On trees and logs they pry off loose flakes of bark. They rarely search amid foliage. In a three-year study in southern Queensland, Brian King learned that each group is a stable association of two to thirteen males and females in about equal numbers, who spontaneously offer food to companions of whatever age, who do not beg for it. The recipient may pass it to another bird, and the latter to a third. In late afternoon, before retiring to rest, the babblers dust-bathe close together, with their bills flicking dry earth over themselves and their neighbors, while they fluff out their plumage and flutter their wings. Then they fly up to a tree and diligently preen their companions as well as themselves, sometimes three simultaneously grooming a fourth. Whether they arrange their feathers or drowse on a perch, they rest in close contact. As daylight fades, the whole group, except the primary female while she incubates eggs or broods nestlings, crowd into a dormitory, of which several are usually available in each territory.

The Gray-crowned Babblers' nest is a bulky, roofed structure, from 12 to 20 inches (30 to 50 centimeters) in external diameter, with a well-enclosed chamber, as much as 8 inches (20 centimeters) wide, entered through a short tunnel under the overhanging roof. The outer walls are composed of twigs up to 18 inches (45 centimeters) long. The inner chamber is made of fine twigs and grasses and lined with a variety of soft materials, including grass, leaf litter, feathers, thistledown, bark fibers, and hair or wool of domestic animals. Whether used for breeding or sleeping, these nests are built by all group members among the outer branches of trees or shrubs, low or high above the ground. In the brood nest the single breeding female lays two or three eggs, which she alone incubates for a period of twenty-one to twenty-five days. While sitting, she is visited by other group members old and young, who usually bring food to her or materials to add to the nest, but often come with empty bill, on a friendly visit, or to learn the state of the nest. All members of the group carry food to the nestlings. When, at the age of about three weeks, the fledglings emerge, they travel with the group, continually calling and begging when an older bird approaches them. In the evening, they join their attendants in the family dwelling.

Nicholas and Elsie Collias found Scaly-feathered Finches on the streets of the city of Bulawayo in Southern Rhodesia (now Zimbabwe) and in the surrounding dry, thorny scrub. Although these finches were not breeding at the time of the Colliases' visit in July, they often saw the birds picking dry grasses from the ground and carrying them to unfinished structures. Occasionally they noticed, in a single small, thorny tree, about half a dozen nests, some old and gray, others fresh and yellow. Made of grass stems and inflorescences, lined with soft, silky spikelets of a different grass, these nests were covered structures, with a thin roof and a side entrance, rarely an opening at both ends.

Watching in the gathering dusk, the Colliases saw up to twelve or fourteen Scaly-feathered Finches enter these dormitory nests. However, most remained empty during the night, causing the observers to ask why so many were built. An explanation was suggested, one evening in Bulawayo, when in the dim light a predatory Fiscal Shrike dashed down upon a nest into which some half-dozen finches had just entered, one by one. The intended victims burst from the nest so quickly that the shrike captured only a beakful of grass. After hiding in the dense foliage of the nest tree for a few minutes, the shrike shot about in a neighboring tree in a futile attempt to catch a finch. Finally, it vanished in the dusk. The finches did not return to the nest where they had narrowly escaped capture but probably retired into one of their other nests. Empty nests may serve not only to mislead predators but also as alternative shelters when the birds are threatened at one of them. Although I find no information as to how fledgling Scaly-feathered Finches sleep, it would be surprising if they did not enter these crowded dormitories with their elders.

The White-browed Sparrow-Weaver, boldly patterned in white, black, and gray, is widely distributed over arid lands of tropical Africa. In Kenya, the Colliases studied these birds in acacia savannas and dry brush; in Zambia, Dale Lewis watched them in light woodland with a fairly closed canopy. In both localities, the sparrow-weavers lived, in pairs or cooperative groups of up to eleven individuals, in clustered, defended territories. Each group contained one monogamous breeding pair and their helpers, mostly their own progeny. They found most of their food on the ground, where they gathered fallen grass seeds or plucked them from growing grasses. They also searched for insects, sometimes using their bills to roll over small stones, clods of earth, or pebbles, or to dig into the soil.

All group members participated in building and maintaining nests throughout the year, but they were stimulated to greater constructive activity by the advent of the rains. They built with dry grasses that they gathered from the ground, picking up loose straws or clipping off standing stems. They also tore from old nests materials for new ones. Situated in trees at no great height, the nests were domed structures about a foot (30 centimeters) long, entered through a spoutlike projection at the end. When one of these nests contained eggs or nestlings, it had only a single entrance. When it served as a dormitory, it was open at both ends, an arrangement that permitted the sleeper to escape through one doorway if a marauder entered through the other. Often two nests, and sometimes three or four, were built in contact, foreshadowing the much closer packing of nest chambers that we shall meet in Sociable Weavers. Each group had two or three times as many nests as members. When not breeding, adults slept singly in these nests.

In a nest with a single doorway, the dominant female of a group of White-browed Sparrow-Weavers laid from one to three eggs, which she alone incubated, sitting for brief intervals alternating with absences of about the same length, so that she covered her eggs for only about half of the daylight hours. When the young hatched, they were brooded only by their mother, who at first supplied most of their food. After a few days, other group members started to feed the nestlings, until most or all of them were so engaged, bringing many small moths and caterpillars. When sixteen or seventeen days old, the one or, less often, two fledglings reared in a nest fluttered out. In the week before a young sparrow-weaver's departure, the adults of its group all joined in building for it a new nest, with a conspicuous doorway to facilitate the entry of the weakly flying fledgling. As night approached, the young bird's mother, sometimes with other group members, would go in and out of the nest while the fledgling perched nearby, guiding and encouraging it to enter. Most often the youngster slept in the nest made for it, but occasionally it entered a different one. If this happened to be the dormitory of an adult, this bird moved to another nest, leaving the fledgling to sleep alone.

Fed at first by all group members, fledgling White-browed Sparrow-Weavers began to find some food for themselves about the second week after their emergence from the nest. In a large group, the mother soon laid again, while the helpers took care of her young. Until they were over three months old and well able to support themselves, juveniles continued to beg for food. During their first six months, they contributed little to nest-building, but at nine months they became active nest helpers. In lodging singly, White-browed Sparrow-Weavers resemble Bananaquits, many woodpeckers, and other birds treated in Chapter 7. However, in preparing for the comfort of their fledglings, they belong with the more advanced dormitory-users. Their clustered nests might be viewed as dwellings with a separate bedroom for each member of the family.

Gray-capped Social-Weavers were studied by Nicholas and Elsie Collias in the semiarid acacia savannas of Kenya. Here the birds lived in groups or families of from three to seven individuals, whose activities centered around a number of nests in an acacia tree, in a colony that sometimes consisted of sixty or more birds. A single small tree might support the nests of up to five groups, each of which occupied a certain part of the branching crown. Two such trees might be little more than 100 feet (30 meters) apart. In small acacia trees, the roofed nests of grasses were well separated; in ant-gall acacias swarming with ants that kept predators aloof, they were closely clustered, with up to eight in contact with each other. Throughout the year, the social-weavers, singly or as many as five together, slept in these nests, which when serving as dormitories were open at both ends, like those of White-browed Sparrow-Weavers. By day, the social-weavers spent most of their time gathering seeds and insects from the ground surrounding the tree where they slept. They did not defend a feeding territory, but at any season might join members of other colonies in a large flock of foragers.

Like other birds of arid regions, Gray-capped Social-Weavers started to breed when the rains returned. Each group had a single reproductive pair. Recently fledged young and older birds from earlier broods helped their parents to build, sometimes as many as six working on one nest. When a nest was chosen for breeding, one of the doorways was closed before the female laid two or three eggs. During the incubation period of about fourteen days, the male parent alternated on the nest with his mate but took much shorter turns than she did. On some nights both parents slept in the nest with the eggs; on others, only one. During the twenty days that the young remained in the nest, they were fed by from two to six attendants, including the parents and helpers both immature and adult. At first the nestlings received mostly insects, but before they fledged grass seeds were added to their fare. The parents, working about equally hard, brought most of the nestlings' meals, but helpers in adult plumage contributed substantially, and juveniles added their mite. Parents and auxiliaries continued to feed the young birds for three or four weeks after they left the nest. Many remained in their natal area for at least nine months, sleeping with their companions in the covered nests with two doorways, and becoming their parents' assistants in the following breeding season.

The greatest of all avian apartment houses are made by Sociable Weavers, sparrow-sized birds clad in streaked brown and buff, with black throats. These monuments to avian industry are conspicuous from afar on the barren landscapes of the parched Kalahari Desert in

Namibia and in neighboring regions of southern Africa, where the preferred site is a stout Camelthorn Acacia with an open crown, although other trees with thick, horizontal boughs, and even telephone poles and the towers of water tanks may support the heavy structures. In warm, wet regions, these nests of perishable vegetable materials would never resist decay long enough to attain their enormous size; but in arid and semiarid southern Africa, the birds add to the same mass year after year. A nest in eastern Transvaal, belonging to a colony reported to be a century old, was an irregular mass about 16 feet long by 12 feet wide (4.8 by 3.6 meters) and several feet thick. Its lower surface was pitted with the entrances to 125 nest chambers. Most nest masses are smaller than this, down to those with less than a dozen chambers. A colony, which may comprise up to five hundred birds, may start several separate nest masses that grow until they coalesce into one. Or a group may occupy two or more neighboring trees.

Working together, colony members start a nest mass by arranging sticks from 4 to 12 inches (10 to 30 centimeters) long in a dome-shaped roof or superstructure above the supporting limb. By inserting dry straws into this mass, where they are held by friction, the Sociable Weavers extend the substructure downward below the branch. Here they make the separate compartments, each of which has a vertical entrance tube up to 10 inches (25 centimeters) long, leading upward to a chamber, about 6 inches (15 centimeters) in diameter, set to one side, so that the whole compartment has somewhat the shape of a chemist's retort. With grass inflorescences, furry leaves, shredded grass blades, and a few feathers, the birds line their chambers softly all around. These rooms do not intercommunicate. The mouth of each entrance tube is surrounded by the ends of stiff, outwardly directed straws in a bristly collar that may repel a human hand and perhaps certain predators. When not breeding, three or four Sociable Weavers often collaborate in the construction of a single compartment, but after breeding begins each pair maintains its own.

Like the nests of Firewood-gatherers and other elaborate structures built by ovenbirds, the Sociable Weavers' homes are kept in good repair only by the unremitting attention of their occupants. Throughout the year, the birds work on them whenever they are not foraging, attending their eggs or young, or resting. Returning from a foraging expedition, most birds bring back a straw, often from a point a mile away. From the ground beneath the nest, they gather fallen pieces and carry them up to reincorporate in the structure. Except soft materials for the chamber's lining, which may be collected until the bird has a billful, a Sociable Weaver carries a single piece, often a straw grasped by its lower end. When only eighty days old, they begin their lifelong occupation of building and repairing.

These massive structures are the permanent dwellings of their builders. At all seasons, the birds sleep in them, each in whatever chamber it wishes in its own section of a great nest mass. As many as five occupy a single room, where they probably rest in layers. At dawn they begin to call in their nests, but usually they do not become active until sunrise or a little later. Then they fly in flocks to their feeding areas, mostly within a mile of their domicile. Spreading over the ground, they advance by hops or a short, shuffling step or two, while they search for the insects that are their principal food, or gather the seeds of grasses. Around ten o'clock in the morning, they return for a long siesta in their nest chambers, protected by the thick roof from the intense desert sunshine through the hottest hours of the day. At about two o'clock in the afternoon, they sally forth again for another spell of foraging. Around sunset, they flock back to sleep in their nest. With certain variations, this is their schedule on warm summer days. On a chilly winter morning, they may linger in their snug quarters much longer, but then their midday rest is often curtailed, and their evening return delayed until after sunset, to compensate for the fewer daylight hours when they can forage.

Like other birds of arid regions, Sociable Weavers have no definite annual breeding period but await showers that in any month may awaken the dormant vegetation with its accompanying profusion of insect life. According to the duration of favorable conditions, their nesting may be brief or protracted; in a prolonged drought, they may fail to breed at all. After a good shower, females lay promptly, in chambers that throughout the year serve as dormitories, usually three or four eggs in a set, with a range of two to six, the number depending upon the abundance of food. Not every room contains a breeding pair. The male and female parents share incubation rather equally, sitting from two to forty minutes at a stretch and keeping their eggs almost constantly covered, for one seldom leaves the nest before its partner arrives to replace it. Occasionally a parent flies out when a third bird enters the chamber, to remain, probably warming the eggs, until one of the parents returns. In thirteen or fourteen days, the eggs hatch. Unlike most passerine birds, Sociable Weavers do not carry away the empty shells but simply drop them through the entrance tube to the ground, where they hardly increase the conspicuousness of the nest mass or the litter beneath it.

Typical passerine nestlings with sparse natal down, newly hatched Sociable Weavers have at the corners of their mouths swollen, creamy white protuberances that probably help the parents to deposit food in the right place in the dimly lighted chamber. At first, the parents brood the nestlings by turns almost continuously. Both bring them insects that increase in size and number as the days pass. Harvester termites that, after a shower in the Kalahari Desert, become abundant and active by day help to nourish nestlings of all ages. Although usually only the parents feed nestlings of first broods, occasionally one or two nonbreeding adults assist them. Another departure from the usual routine of passerine birds is the parents' failure to carry away droppings, which are not enclosed in gelatinous sacs to facilitate their removal. From the day they hatch, nestlings void their wastes over the entrance tube, through which they fall to the ground.

After leaving the nest at the age of twenty-one to twenty-four days, fledglings return to rest and sleep in it. They might not find the entrance to their own chamber amid so many others if the parents, who recognize them individually amid the throng, did not guide them to it. The fledglings also recognize their parents, whom they approach calling for food with pleading postures. They beg less intensely from other adults; but only parents were seen to feed their fledglings, continuing this for about a week after their first departure.

In an interval of favorable weather, Sociable Weavers may raise up to four broods in nine months of almost continuous reproduction. As soon as the nestlings of the second brood hatch, young of the first brood, when they are only twenty-five to thirty days old, help their parents to feed them. With each successive brood, the number of young helpers increases; a fourth brood was attended by eleven birds, including the parents and nine juveniles from three earlier broods. After acquiring adult plumage when about four months old, most young remain to breed in their natal colony. In contrast to many other cooperative breeders, some of which require from four to six years to reach reproductive maturity, and despite the protection afforded by their enduring nests, Sociable Weavers start to breed at an early age and die young. Gordon Maclean, upon whose painstaking study the foregoing account is largely based, estimated that they rarely survive for more than three years and a few months.

The estrildid finches are a family (Estrildidae) of 137 species of waxbills, grassfinches, and mannikins (not to be confused with the tropical American manakins) that range from Africa through southern Asia to Australia. The brilliant colors of many of these small, mostly thick-billed birds, easily reared in captivity, make them favorites of aviculturists. Most are highly gregarious and flock over grasslands and semi-open country; a few inhabit forests. Their food consists largely of ripening or ripe grass seeds, with an admixture of insects. With grasses they build covered nests with a side entrance, often a long tunnel. On the roof of the nest of many African species the male builds an open or incomplete nest, possibly as an outlet for excess energy, as no other use has been discovered. Moreover, he helps his mate to incubate the four to six (rarely more) white eggs, which often hatch in the unusually short period of ten days but in some species require two weeks of incubation.

Nestling estrildids are remarkable for the elaborate patterns on their mouths and tongues; for the glistening, light-reflecting knobs at the corners of their mouths, which guide the parents to them in the dimly lighted nest; and for their nestling period of two or three weeks, which is exceptionally long for such small passerine birds. Unlike most passerines, the parents fail to remove their nestlings' droppings, which accumulate in the nest, but in the arid regions where many of these birds live they dry so quickly that they do not soil the plumage of the young birds or their attendants.

Many estrildid finches sleep in a dormitory, which may be a breeding nest, one constructed specially for this purpose, or the abandoned nest of some other bird. In Australia, Klaus Immelmann learned that in the evening a parent Zebra Finch, most often the father, leads newly emerged fledglings to a nest, where the parents feed them after their entry. Soon the young enter spontaneously, and they continue to lodge there until they attain independence. Usually the parents sleep in the old nest with their young. However, their sleeping arrangements differ greatly with the condition of this nest. If it has become too small for the growing family, too soiled with droppings, or too dilapidated with long use, they seek some other nest for a dormitory, where, according to its capacity, the fledglings sleep alone or with their parents. Once a family occupied two nests, in each of which one parent slept with part of the brood. For a dormitory, the male parent repairs or remodels, often before the young fledge, an empty nest built by the Zebra Finches or some other bird. When parents and fledglings sleep apart, the former install the latter in a dormitory before they enter their own.

In Kenya, V. G. L. Van Someren found that Black-headed Manikins sleep in old brood nests or weavers' nests. If one of these is not available, several individuals join in building a dormitory large enough to hold a number of them; eleven adults slept in one that he examined. He also ascertained that fledgling Crimson-cheeked Blue

Waxbills returned to sleep in their nest. In this species, as in the Barred Waxbill, Bronze-headed Mannikin, and probably other estrildid finches, both parents pass the night in the nest during incubation.

References: Collias and Collias 1964; 1978; 1980; Emlen and Demong 1984; Fraga 1979; Haverschmidt 1951; 1953; Hudson 1920; Immelmann 1962; B. R. King 1980; Lewis 1982; Ligon and Ligon 1979; Maclean 1973; MacRoberts and MacRoberts 1976; Moreau and Moreau 1937; Narosky, Fraga, and de la Peña 1983; Norris 1958; Schifter 197?; Short 1973; Short and Horne 1984; Skutch 1935; 1944; 1948a; 1948b; 1958; 1960; 1969a; 1969b; 1980; 1983a; 1987; Skutch and Gardner 1985; Van Someren 1956; C. Vernon, letter to author.

ELEVEN

Sleeping on Cold Nights

In the preceding chapters, we surveyed birds that sleep in dormitories singly, in pairs, in families of parents and dependent young, and, finally, in groups of closely cooperating individuals, mostly a mated pair with their progeny, plus at times recruits from other families. A number of facts suggest that for these birds the dormitory is a luxury rather than a vital necessity. They are mostly birds of milder climates where prolonged low temperatures are absent. Most of their neighbors of different species, including those with similar diets, survive very well with no more nocturnal shelter than foliage provides. Some of the dormitory-users, like the Yellow-olive Flycatcher, do not occupy dormitories after the breeding nest in which they have been lodging deteriorates. A number of them, including Blue-throated Green Motmots, Hairy Woodpeckers, Red-crowned Woodpeckers, Bananaquits, and others, leave their fledglings outside while the parents continue to sleep in their shelters. The immature birds are doubtless less hardy than their elders, yet they do not appear to suffer severely from exposure.

Probably, in many cases, the dormitory is more important as protection from predators than from the weather. The Fiscal Shrike that tried to capture Scaly-feathered Finches who had just retired for the night, as related in the preceding chapter, would probably, had they been roosting in the open, have seized one of them instead of a beakful of grass. Often I have found a covered nest, emptied of its contents, with a round hole in the top or back, apparently made by a predatory bat or other small mammal that could not find the doorway. An adult sleeping in such a nest would certainly have had time to escape while the marauder was making the hole. All the evidence points to the conclusion that covered nests and holes of various sorts were first made or found to protect eggs, nestlings, and incubating or brooding parents from tropical sunshine, tropical downpours, and predators, and that, because they also offered comfortable lodging for adults and fledged young, they were secondarily used as dormitories. By no means all birds that breed in covered nests sleep in such structures. In the case of cooperative breeders who are also dormitory-users, sleeping in close contact, like mutual preening and exchanges of food, helps to bind the cooperators more firmly together.

In all the foregoing examples of dormitory-using, we rarely found more than about a dozen birds sleeping in the same nest, or the same compartment of an avian apartment house, and usually the occupants were fewer. In the present chapter we shall meet much larger aggregations of sleeping birds, mostly strangers drawn together by force of circumstances rather than members of the same family or cooperative group. Accordingly, despite the number of birds involved, this eighth stage does not represent a further advance in the evolution of the dormitory habit so much as a special development, an emergency situation imposed upon the birds by the vital need to keep warm and conserve energy when the weather turns severe and food is scarce. Many observations indicate that these large aggregations are not the participating birds' preferred way of sleeping but forced departures from their normal habits. And, as we shall see, they bring special perils.

Let us begin with the Winter Wren, the single member of its family that is distributed around the world, mostly at fairly high northern latitudes. In winter it has been found sleeping singly in the same diversity of situations that males occupy in summer: old nests of Wrens, House Sparrows, House Martins, Barn Swallows, and Long-tailed Tits, as well as cup-shaped nests of Eurasian Blackbirds and other species amid sheltering vegetation, likewise in thatched roofs, haystacks, crevices in old walls, and among aged ivy stems. At other times, Winter Wrens huddle together in large numbers: in Britain, nine slept in an old nest of the Song Thrush, ten squeezed into a coconut shell, and forty-six into a nesting box; in the United States, five perched side by side on a branch beneath a canopy of accumulated dead pine needles, pressed so closely together that their feathers intermingled, and thirty-one took refuge in a bird box 6 inches (15 centimeters) square. In such crowded quarters the sleepers squat upon each other's backs, often in two or three layers or tiers, with heads inward. When disturbed, or even caught and released, they may promptly return to the same dormitory, although, in similar circumstances, most birds would seek another place to pass the remainder of the night. Why these striking contrasts in behavior?

Tree-Creepers cuddled together on rough bark

Investigating this problem in England, Edward A. Armstrong concluded that "high humidity and low temperature in association are crucial in eliciting and maintaining social roosting" in Winter Wrens. That these little birds clumped together of necessity rather than from preference was clearly demonstrated by the behavior of a banded wren, Silver, who, in the changeable weather of January 1945, alternated between his solitary dormitory and one shared with others. On the milder nights, he slept alone. Long before Armstrong correlated this wren's variable behavior with meteorological records, it had been noticed by country people, who summarized it in a jingle:

> When tom-tits cluster
> Soon there'll be a bluster.

From as much as a mile (1.6 kilometers) away, the wrens may gather on a cold evening to sleep with companions. They appear to follow a leader who knows a snug nook and moves purposefully. Since the larger the company the more warmly they will sleep, with songs and calls potential bedfellows attract one another. When driven from their dormitory, they return promptly because to sleep dispersedly on a freezing night may be lethal to them.

Eastern Bluebirds also change their way of sleeping when the weather turns very cold. In Arkansas, Ruth Thomas found that in winter resident birds snuggled, three or four together, in clusters of dead oak leaves on low-hanging boughs, near their nest boxes. Migrating flocks roosted in the same way. Only in the coldest weather did the bluebirds sleep in boxes. One January, during a snowy week when the thermometer fell to 5° F. (−15° C.) two pairs who had fought for possession of a box huddled together in it.

Over much of the northern hemisphere, typical magpies roost gregariously in trees and shrubs. Occasionally, however, they sleep in their nests, ample bowls of clay or mud bound together with vegetable materials, built upon a foundation of coarse sticks, which also roof them over, leaving an opening in the side. Magpies are most likely to sleep in nests in the extremely cold winters at high altitudes in northern lands, as at 7,200 feet (2,200 meters) in Wyoming, where Michael Erpino found Black-billed Magpies lodging in their old nests, and at higher altitudes in Tibet, where E. Schäfer learned that up to ten magpies regularly slept together in large dormitories, one of which was a honeycomblike structure of sticks, suggestive of several small magpie nests built in contact. Where unoccupied magpie nests are available, many small passerines take refuge in them during storms of hail or rain. Old magpie nests are also occupied for breeding by Great Horned Owls, Mallards, hawks, and some passerine birds.

As the light of an autumn day faded in the Rocky Mountains of Colorado, Owen Knorr watched a chattering flock of Pygmy Nuthatches approach and enter the high, hollow stub of a large pine tree. He estimated that at least one hundred birds entered one of the several cavities in this stub, with fifty more in others. From similar hollows in other pine stubs, a number of dead Pygmy Nuthatches, all adults in a remarkable state of preservation, were recovered, thirteen from one cavity. In the much milder climate of coastal central California, Pygmy Nuthatches continue through the winter to sleep in family groups, or perhaps two families may occupy the same hole, with a maximum of only eleven birds, as told in Chapter 10. White-breasted Nuthatches, who usually sleep alone, seek bedmates in severe weather. On a winter evening, twenty-nine, arriving separately rather than in a flock, vanished into a large crack in the trunk of an old dead pine tree.

After Wellingtonia trees introduced into England grew large enough to have furrowed bark, European Tree-Creepers discovered that they could dig little hollows into the soft, fibrous cortex of this conifer, in which they slept singly, clinging upright with back and tail exposed. In winter in Germany, about fifteen of these little brown birds clustered on the rough bark of a tree in a feathery ball, tails sticking out. In winter and early spring in the United States, Brown Creepers also sleep together for warmth. Two snuggled tightly in a nook between a nest of mud-dauber wasps projecting from a chimney and the adjoining wall of the house, close under a roof. Eleven slept nightly in a hollow beam protruding from the side of a barn.

Like other titmice, Blue Tits nest in a cavity of some sort, but as a rule they roost in evergreens or ivy on a wall. In hard weather they seek the more adequate shelter of a hole, or whatever protection they can find. In the winter of 1916–1917, when many perished, Blue Tits slept in a flowerpot hung against a wall, an empty coconut, a broken bed bottle, and a glass wine bottle. Years later, in Len Howard's home, so exceptionally hospitable to birds, Blue Tits slept not only in crevices on the outside of the cottage but indoors in cardboard boxes that she provided for them, or in any suitable nook they could find in her walls or furnishings. Great Tits also took full advantage of the shelter offered by her rooms.

In summer in the Netherlands, Great Tits roost well concealed in the crowns of deciduous trees, but in autumn, when the trees shed their foliage, they move to cavities in trees or bird boxes provided for them, always insisting upon sleeping alone. In a tract of woodland with many boxes, acquisitive individuals took possession of several, visiting them by day and, when necessary, fighting other claimants, most fiercely just before they retired in the evening. Probably they wanted several boxes so that, if disturbed at one, they might move to another. As a consequence of this acquisitiveness, coupled with the dominance of males over females and of adults over young, a larger proportion of resident males than of females, and of adults than of juveniles, enjoyed the protection of the nest boxes through the coldest months. However, as the breeding season approached in March, and females became ascendant over males, they became the more numerous occupants of the boxes. H. N. Kluyver, to whom we owe this study, found evidence that sleeping in boxes contributed to the greater survival of male Great Tits than of females.

In general, titmice sleep in dormitories less consistently than do certain other birds that breed in holes or covered nests, such as woodpeckers and wrens. They appear to seek enclosed spaces for sleeping chiefly in cold weather. In Ireland, Coal Tits sleep in holes in decaying trees, usually pecked out by themselves, or in the dense cover offered by ivy. In the United States, Black-capped Chickadees usually roost amid foliage but occasionally sleep in a cavity in a tree or an old open nest of some other bird. In autumn and winter in Tennessee, Carolina Chickadees slept singly in small "natural" cavities in trunks and snags. They fouled their bedrooms with their droppings and, probably for this reason, changed them frequently. On cold nights, one lodged in a gourd. In the eastern United States, Tufted Titmice spend winter nights in old woodpeckers' holes, cavities in posts and stubs, and bird boxes. They often rest in their shelters on dull, drab days. In the West, Plain Titmice sleep in holes or amid foliage that clusters around them, simulating a cavity. In California, Chestnut-backed Chickadees roosted, even in summer, on a thick wire cable beneath the eaves of a house, and on the petiole of a leaf of English ivy amid its concealing foliage, close beneath a roof. In Chapter 3, I told how in Siberia Willow Tits passed winter nights in disused rodents' burrows beneath snow. Female titmice often begin to sleep in their nest chambers before they lay. A Plain Titmouse in California did so for a month.

Both parent Long-tailed Tits sleep in their cozy, feather-padded nest with eggs or nestlings. After breeding, they wander through hedgerows, coppices, and woodland in chattering flocks of half a dozen to about thirty individuals. In milder weather they roost, pressed together in a row, low amid shrubbery or on the bare branch of a tree. In severely cold weather, they sleep in a dense thicket or in a hole, cuddled together in a compact ball, their long tails sticking out at diverse angles. Like their relatives, the Long-tailed Tits, Bushtits sleep in their nest only while breeding, after which they roost in trees. In the state of Washington, Susan Smith learned that on warmer nights in winter they rest with a short interval between individuals, but freezing temperatures impel them to huddle in contact through the long hours of darkness.

When not nesting, most members of the predominantly tropical parrot family, except some of the smallest species, roost amid foliage. In the severe winters of eastern United States, the now extinct Carolina Parakeets slept gregariously in old holes of the larger woodpeckers or in hollow trees, where they clung upright, suspended from their bills, with additional support from their feet. On a cold winter day, hundreds of these birds imperfectly adapted to a northern climate were found in the hollow trunk of a large sycamore tree, too torpid to fly or make a move to escape when the tree was cut down. When taken into a house, some of these parakeets revived and enjoyed fruits and nuts offered to them. On the opposite side of the tropics, in Argentina, Monk Parakeets also need protection from cold weather, which they find inside their large apartment houses built of interlaced sticks. Another member of the family that reaches fairly high southern latitudes, including stormy Patagonia, is the Burrowing, or Austral, Parrot. In vertical cliffs these parrots dig for their nests crooked, sometimes intercommunicating burrows up to 10 feet (3 meters) long. In the evening they approach the burrows in spectacular flights to sleep in them. The southernmost population migrates northward in winter, when in a more benign climate the birds apparently roost in trees.

In spring of the northern hemisphere, a sudden cold spell, bringing snow and ice, sometimes afflicts migratory birds that have prematurely arrived from the tropics; or, in autumn, a cold wave overtakes those who have too long delayed their southward departure. In such weather, thousands may perish. Mortality is especially high among aerial flycatchers such as swifts and swallows, for few insects fly in chilling air, and just when these birds need more food to maintain their body temperature, they find little. In such circumstances, they cluster or clump together to keep warm.

In North America, in both spring and fall, migrating Chimney Swifts pass the night in tall, inactive chimneys or warm air flues of factories and public buildings, which have largely replaced the big, hollow trees where they formerly lodged. High in the air, massed thousands of swifts swirl in spectacular circles or more complex figures before they stream downward into the chimneys. Here they cling to the wall in a continuous black sheet, each, where possible,

*Chimney Swifts funneling into chimney and (*inset*) clinging to wall inside*

resting its head upon the base of the tail of the bird above it, with the ends of the wings of the upper bird on the back of the bird below. Usually they do not rest in layers, some wholly above others; but when they have more difficulty keeping warm, they pack more solidly together. On a dull September day, more than a hundred of them clung, two or three deep, to the bark of a tall oak tree, in a patch nearly 5 feet (1.5 meters) high and about 8 inches (20 centimeters) wide. All but a few of the uppermost kept their heads hidden in the feathered mass. Here the crowd remained through the night, and some stayed until the following afternoon. Even in favorable weather, the Chimney Swifts' morning departure from a chimney is gradual and may continue until past the middle of the forenoon, while many of the swifts who have flown out at dawn re-enter around sunrise. When the temperature approaches the freezing point, the swifts remain in the chimney longer and return after brief, probably poorly rewarded excursions for foraging.

In Europe, prolonged cold, tempestuous weather, even in midsummer, makes Common Swifts cluster in similar fashion. After three days of continuous cold rain, swifts at Konstanz in Germany huddled together in a mass more than 10 yards long and 1 or 2 yards wide (10 by 1–2 meters). Another clump in the same locality contained two hundred birds. In these large aggregations, individuals on the outside try to force their way deeper into the mass, where they would be warmer. Probably it is those in the more exposed positions who succumb during the night and are found dead next morning; in the cluster of two hundred swifts at Konstanz, twenty perished. When weak from cold and hunger, Common Swifts often seek shelter in buildings. In England, one entered a school dormitory through a window and crept under the pillow of an occupied bed. On an afternoon in May, when swifts migrating northward through southern France encountered a strong headwind, one entered a room through a window and was put outside. When this or another swift intruded in the same way, the window was closed. After this, a swift clung to the outside wall of the house, against a corner of the window, where it was joined by nineteen others, each touching its neighbors. Here they remained through the night.

Not only in cold weather do swifts sleep in large companies. In a huge hollow sycamore tree in Kentucky, Audubon found about nine thousand Chimney Swifts, mostly adult males with a few juveniles from early broods, sleeping as early as July. On the tropical island of Trinidad, David W. Snow found Short-tailed Swifts lodging in manholes of an abandoned drainage system. Outside the breeding season, thirty or more slept in a clump on the concrete wall. Molting swifts, to the number of fifty or more, remained in these manholes on fine sunny mornings. In El Salvador, Joe T. Marshall, Jr., watched a flock of about fifty of the large White-collared Swifts enter at nightfall a fissure behind a waterfall in a deep canyon. In the fading light, single birds or small groups fluttered through the spray into the cleft. At sunset in August in California, Gayle Pickwell watched from one to two hundred White-throated Swifts stream at great speed through a narrow crevice in a rocky canyon wall.

When perching in a long row on an overhead wire, swallows are often evenly spaced, a distance of about the birds' own length separating closest neighbors. Likewise when resting in trees, swallows of various species preserve their "individual distance." But when the temperature drops low, swallows sometimes close ranks, perching much nearer each other than they usually do, frequently in contact, with sometimes a third individual trying to squeeze between two that had huddled together for warmth. Even swallows of different species may press against each other. On a cloudy, chilly morning in late May in Massachusetts, forty Bank Swallows rested in a compact row on telephone wires, with huddled groups of two, three, or four birds near them. Pressed against some of these Bank Swallows were a few Barn Swallows and Tree Swallows. Despite the chill, a few of these swallows remained aloof. Similarly, male House Sparrows, who even in moderately cold weather keep an interval between themselves and their nearest neighbors, perch closer together or in contact when the temperature falls well below the freezing point, as Kathleen Beal demonstrated with captive birds.

When they find a protecting cavity, swallows weak from cold and hunger crowd into it, regardless of risks. In Manitoba in mid-May, when northerly winds brought freezing rain, snow, or hail, eight Tree Swallows were found, stacked one on top of another, in the small hole of a Downy Woodpecker. When removed, only two at the bottom of the heap were still alive, and these were so weak that only one survived. Had they not been rescued, both would probably have succumbed because they lacked strength to push up past the corpses of their companions above them. The dead swallows had lost about a quarter of their normal weight; their empty stomachs revealed that they had starved. After the same storm, nine dead Tree Swallows were removed from a nest box. In an earlier May, eleven corpses were found in a nest box in New York.

Tree Swallows winter in southern Florida, where the subtropical climate is at intervals invaded by freezing north winds, disastrous to introduced tropical plants, birds, and even small fishes. In such a cold spell at the end of January 1940, when large numbers of dead and dying Tree Swallows lay scattered over the ground, survivors clustered like bees in sunny spots, such as a seawall. Of fifty that had crowded into an old Pileated Woodpecker's hole, twenty-eight were dead when found, or died soon after removal. A few promptly flew away, or revived in the sunshine. The stomachs of the dead birds were empty; the plumage of many was soiled with white, lime-laden excrement.

During the stormy May in Manitoba mentioned above, Patrick Weatherhead and his associates found in a garage eight Barn Swallows stacked at least two deep in an old Barn Swallow's nest. With heads inward, they called incessantly, while the swallows on top appeared to be trying to force their way beneath those below them. Next morning, all the swallows had gone from this nest, except one that lay dead in it. A search in seven other buildings in the neighborhood produced five dead Barn Swallows, three of them in nests and two directly below the nests. In both the United States and England, Bank Swallows, or Sand Martins, perished in numbers in their unfinished burrows where they had sought refuge during late snowstorms; fifteen corpses were removed from one tunnel.

These observations on swallows, with those on Pygmy Nuthatches that died in hollow pine trunks, remind us that the expedient of clustering tightly in closed spaces is not without special dangers. Birds deep in the feathered mass may suffocate; those on the outside may squander the last remnant of their depleted sources of energy by trying to push inward where they would be warmer; the plumage of some may be soiled by the excreta of their companions, reducing its capacity to retain heat; survivors near the bottom of the heap, lacking strength to push upward, may find themselves buried alive by the corpses above them. Clustering may be the last forlorn hope of birds dying of cold and hunger.

The foregoing remarks hardly apply to birds of warm lands who sleep in clusters, whether in the open, like wood-swallows, mousebirds, and hanging parrots, or in dormitories, like Banded-backed Wrens, Gray-crowned Babblers, and many others. In the first place, they will usually be well fed when they go to rest, with strength to shift their positions if they find breathing difficult. The clustered birds are rarely so numerous as those that sometimes huddle together in cold weather. If the dormitory becomes uncomfortably crowded, groups often have alternative lodges to which some of their members can move. Birds of warm lands appear to bunch together at night for comfort or companionship rather than dire necessity. Perhaps, too, they sleep in clusters for greater safety, because a wakeful individual may detect the approach of a marauder and warn its companions, or the clump may "explode," alarming the enemy.

As told by Irven Buss, a farmer in Deerfield, Wisconsin, developed procedures whereby he increased the number of Cliff Swallows nesting on his unpainted barn from a single pair to a colony of over four thousand birds in thirty-eight years. Cory Bodeman, the farmer, learned that both nesting parents and their young survived much better when he removed the old swallows' nests, occupied by House Sparrows with their attendant insect vermin, before the swallows arrived in the spring, so that all were obliged to build clean, new clay-walled urns for themselves. This was wholly beneficial when the spring was mild; but if the temperature suddenly dropped after the birds' return from the south, great numbers might perish from exposure when deprived of their usual shelters. As a compromise between the swallows' need of their old nests to keep warm and the advantages of forcing them to build new ones free of vermin, Bodeman knocked only some of the old nests from the barn wall before the birds' appearance in the spring; then, after the danger of cold weather had passed and the swallows started to build, he removed the remainder of the old structures.

Cliff Swallows and House Sparrows are not the only birds that seek shelter from adverse weather in well-enclosed nests. On cliffs beside the Snake River in the state of Washington, William Shaw watched Rosy Finches retire on winter evenings into swallows' nests clustered in depressions in the irregularly fractured face of basalt rock, while the builders of these nests were far away in warmer lands. He did not tell us how many finches occupied a single urn. In summer, these hardy finches nest in holes and fissures in vertical cliffs above timberline on high mountains of western North America, amid glaciers and lingering patches of snow. In autumn they descend to open fields and bushy areas at much lower elevations, where they visit feeders or windowsills for seeds set out for them. They sleep gregariously in clefts in rocks, old mine shafts, deep wells, strings of standing freight cars, and buildings where they find overhead shelter and protection from wind, as well as in Cliff Swallows' nests. Winter after winter, they return to the same building, elevated water tank, or substructure of a pier on the Great Salt Lake.

Birds of many other kinds, including those that usually neither nest nor sleep in closed spaces, are quick to take advantage of whatever shelter they can find when subfreezing temperatures strike them. During the same January when so many Tree Swallows succumbed to cold and scarcity in Florida, many Yellow-rumped, or

Myrtle, Warblers, one of the hardiest of the wood warblers, left their frozen corpses on the ground. Some sought shelter in an abandoned outbuilding amid the scrub, where the bodies of fifteen were later found. More fortunate was a European Robin who on winter nights roosted in a eucalyptus tree on Edward Armstrong's porch. He learned to enter before the door was closed in the evening, and he waited patiently until it was opened in the morning and he could fly out, never beating himself against the window panes, in the manner of other birds who blundered into the porch. In Tennessee in winter, a pair of Northern Mockingbirds slept on separate perches in a garage, two passed the night in porches, and one dropped down a chimney at dusk. Others roosted in evergreen trees and shrubs. In Chapter 9, I told of a Carolina Wren who slept in a conservatory on winter nights.

As related in Chapter 5, with the exception of Blue-throated Green Motmots, all the birds that I have found sleeping in dormitories on forested Central American mountains, up to about 10,000 feet (3,050 meters) above sea level, belonged to families whose members also occupy such shelters in warm lowlands. The largest number of birds that I found clustering together in a cavity was sixteen Prong-billed Barbets, at an altitude of 5,500 feet (1,675 meters) in Costa Rica. These small, stout birds might have huddled together in a shallow tree hole for warmth, for nights could become chilly, although freezing temperatures were not recorded at the altitude where the barbets slept; but I believe that the primary reason for their gregarious roosting was shortage of suitable cavities. As far as I saw, the barbets carved new holes only in the breeding season. As these holes fell or deteriorated in the long, rainy months after the young fledged, the barbets solved their housing problem by lodging in larger companies.

Between 8,000 and 10,000 feet (2,440 and 3,050 meters) in Guatemala, radiation into cloudless nocturnal skies during the dry season, which there coincides with the northern winter, whitened open fields with frost. Nevertheless, I did not find frost or ice in the woods or beneath sheltering trees or shrubs. Although the vegetation in general was not frost-hardy, I noticed the effects of freezing only in low plants exposed to the open sky, or in hollows where air chilled on exposed slopes had settled. Here, where the temperature of the air in woods and thickets remained at least a few degrees above the freezing point, most of the birds appeared to find adequate shelter while roosting amid foliage, unless they were of families that breed and sleep in nests or cavities of some sort, wherever they may be.

Above timberline on high tropical mountains, the climate becomes much more rigorous, with cold rain, hail, snow, chilling winds, and nocturnal temperatures well below the freezing point, especially when cloudless dry-season nights coincide with winter some degrees north or south of the equator. Here birds may resort to ways of sleeping unknown among their relatives of milder climates. Most hummingbirds roost on trees and shrubs, sometimes wholly exposed to the sky, as we noticed in the case of the Long-billed Starthroat in Chapter 4. High above timberline on the bleak, treeless puna of Peru, up to 13,000 feet (4,000 meters) or more, the Andean Hillstar fastens its thick-walled, open nests to the fractured rock of vertical cliffs, or to the walls and roofs of caves and mine shafts, much more frequently than it builds them on plants.

With feet and toenails exceptionally large and strong for a hummingbird, the hillstar sleeps clinging, head up, to vertical rocks or walls, using its tail for support, much as woodpeckers do. Or, with body horizontal and back downward, it clings to the underside of an overhanging rock or the roof of a cave. Less often it perches upon a horizontal surface. In southern Peru, where even at tropical latitudes the influence of the austral winter is felt, hillstars seek more protected situations in winter than in summer. In the colder season they sleep in caves and mine shafts, where often they press into crevices and holes in the rock barely large enough to contain them, or they cling to the interior walls of deserted buildings. In these situations, the nocturnal temperature rarely falls to the freezing point, even when the air outside is 27° F. (15° C.) lower.

Caves occupied by hillstars, or similar ones, also shelter a wide variety of birds, including the Andean Tit-Spinetail and Stout-billed Cinclodes (both ovenbirds), Black-bellied Shrike-Tyrant, Rufous-naped Ground-Tyrant, Cinerous Ground-Tyrant, and Black-fronted Ground-Tyrant (four American flycatchers), Gray-hooded Sierra-Finch, American Kestrel, and Great Horned Owl.

Prominent on the wide, barren landscapes of the Peruvian puna, around 13,000 feet (4,000 meters) above sea level, stand, isolated or loosely clumped, stout plants of the bromeliad *Puya raimondii*. From the summit of a thick, leafy trunk about 12 feet (4 meters) tall, these gigantic relatives of the pineapple send up a single huge inflorescence about 20 feet (6 meters) high, densely covered with myriad greenish white flowers. The margins of the long, broad leaves are armed with formidable, inwardly curved, sharp spines, each about 6 inches (15 centimeters) long. No other plant of the treeless puna offers birds such good protection from predators as well as from the rigors of the climate; many take advantage of it, for nesting, sleeping, and daytime shelter in inclement weather. Black-winged Ground-Doves forage in flocks of twenty or more amid the grass around the puya and sleep on its leaves, as far inward toward the stem as they can go, sheltered from rain and hail. They rest with their bodies parallel to the leaves and heads outward, sometimes as many as thirty on a single plant. Thick deposits of their droppings cover the grooved leaves. Rufous-collared Sparrows, Ash-breasted Sierra-Finches, and Gray-hooded Sierra-Finches also find protection amid the puya leaves. The Andean Hillstar sometimes attaches its nests to the spiny margins of dead, drooping leaves.

Although most pigeons roost amid vegetation, in the high Peruvian Andes above timberline, Bare-faced Ground-Doves lodge in holes in walls. Massed together beneath great rocks sleep Ash-breasted Sierra-Finches, Gray-hooded Sierra-Finches, Plumbeous Sierra-Finches, Rufous-collared Sparrows, Bright-rumped Yellow-Finches, and Bar-winged Cinclodes. In Bolivia, at the extremely high altitude of 17,400 feet (5,300 meters), more than two thousand White-winged Diuca-Finches crowded together for warmth in a crevice in a glacier. When we add to these the Speckled Teals, Andean Flickers, ground-flycatchers, and ovenbirds that in the high Andes nest or sleep in holes and burrows, it becomes clear that in the thin, cold air of these altitudes a large proportion of the birds seek relief from the nocturnal chill in enclosed spaces.

High on Mount Kenya in equatorial Africa, the giant Tree Groundsel of the composite family has a growth form not unlike that of the Andean puya. In the matted clusters of dead leaves that drape around the upright trunk, the Mountain Chat excavates deep holes, into which Scarlet-tufted Malachite Sunbirds retire to sleep through bitterly cold nights. One evening, two females and a male entered the same cavity. Even in the daytime, immature sunbirds sought warmth and repose in these snug niches.

References: Armstrong 1940; 1955; Beal 1978; Bent 1940; 1946; 1948; Buss 1942; Carpenter 1976; Christy 1940; Coward 1928; Dixon 1949; Dorst 1956; 1957a; 1957b; Edson 1943; Erpino 1968; French 1959; French and Hodges 1959; Goodwin 1976; Groskin 1945; Grubb 1973; Howard 1952; Kluyver 1957; Knorr 1957; Lack 1956; Linsdale 1937; Marshall 1943; Meservey and Kraus 1976; Norris 1958; Pearson 1953; Phillips and Black 1956; Pickens 1935; Pitts 1976; Ruff 1940; Ruttledge 1946; Shaw 1936; Skutch 1960; Smith 1972; Snow 1962; Stone 1950; Tennessee Ornithological Society 1943; R. H. Thomas 1946; Weatherhead, Sealy, and Barclay 1985; Welter 1935; J. G. Williams 1951; L. Williams 1941; 1947.

TWELVE

How Sleeping Birds Conserve Energy

Few of the facts of natural history are more wonderful than the ability of a bird weighing much less than an ounce, or only a few grams, to preserve a temperature of about 104° F. (40° C.) through a long winter night when the surrounding air is far below the freezing point and snow blankets fields and woods. This amazing ability appears the more improbable when we remember that the fuel to keep the bird's vital flame burning must be gathered, in the form of dormant insects or frozen berries, by ceaseless searching throughout a short, bleak winter day. In such circumstances, the small bird's feather garment appears hardly adequate for insulation; with far more massive bodies, we ourselves are often cold under layers of thick blankets.

Despite its efficiency as a producer of heat, the avian organism cannot escape the law of the conservation of energy; it can keep the flame alive only as long as the fuel lasts. In short days when food is scarce, the bird may not succeed in gathering enough to sustain it through the long hours of darkness that follow. For birds that remain at middle and high latitudes through cold winters, any arrangement, anatomical, physiological, or behavioral, that reduces heat loss will have great survival value and be promoted by natural selection.

Heat, a form of energy, is transferred from body to body by conduction, convection, and radiation. When two bodies at different temperatures are in contact, heat is conducted directly from the warmer to the cooler. Convection might be regarded as an intensified form of conduction: when cold water or air flows over a warm body, the heat conducted to it is promptly carried away, maintaining the temperature gradient that sucks ever more warmth from the body. Radiation conveys energy between bodies separated by still air or empty space.

A bird's chief defense against loss of heat is its garment of feathers, supplemented by subcutaneous fat when this is present in appreciable quantity. The number of feathers counted on birds ranges from 940 on a tiny Ruby-throated Hummingbird to 25,216 on a Whistling Swan, 80 percent of whose feathers were on its head and neck. The swan had only 4 feathers for each gram ($\frac{1}{28}$ ounce) of its body weight, whereas the hummingbird bore 335 feathers per gram of body weight. It needed more not only because its feathers were much smaller but also because its body's surface, through which heat is dissipated, was much greater in proportion to the mass of its body, where heat is generated, than in the case of the swan. Among small passerines of eastern North America, the number of feathers on a bird ranges from 1,200 to 2,500, occasionally more or less. A Bald Eagle that weighed about 9 pounds (4,082 grams) bore 7,182 contour feathers that weighed 20 ounces (586 grams), or 14 percent of the bird's total weight. The eagle's feathers weighed more than twice as much as its skeleton—an indication of their importance to a bird. Verdins, Black-capped Chickadees, some Old World titmice, and Golden-crowned Kinglets wear feathers that weigh about 11 percent of their body weights, but most small passerines have relatively less plumage. Just as some mammals have a thicker coat of fur in winter than in summer, so some birds wear more feathers in the colder season, when they are most needed. Certain House Sparrows had 11.5 percent fewer feathers in the warmer months.

Each feather is connected to a system of muscles that enables the bird to raise or lower it. By fluffing out its plumage, so that it contains more tiny air spaces, the bird diminishes its conductivity of heat. By compressing or sleeking down its feathers, it increases conductivity, which may help to prevent overheating in hot weather. By turning back its head and burying it amid the plumage of its shoulders, the bird not only decreases the total heat-dissipating area of its body but, in species with bare facial skin, a naked comb, or wattles, covers surfaces which can lose much heat. In domestic fowls, this sleeping posture may reduce heat loss by as much as 12 percent.

The lower legs and toes of many birds of cold climates, featherless except in a few species, would dissipate much energy were it not for the arrangement of the blood vessels in the upper leg. Venous blood that has lost heat in the lower extremities passes on its upward course through a network of tiny arteries carrying warm blood downward from the heart. The arterial blood delivers much of its heat to the stream that returns it to the body, so that less is borne downward to be lost through the feet. This heat exchange is especially important to birds that swim in cold water or stand on ice, snow, or frozen ground. It is so efficient that a gull standing in icy water may lose only 1.5 percent of its metabolic heat through its feet. Nevertheless, a bird must keep its feet from freezing, at whatever cost in energy. By resting or sleeping on one foot and raising the other against the body, as many birds do, not only in cold weather but often when it is not so cold, loss of heat is further reduced. On cold days, Sandhill Cranes have been seen flying with both legs drawn up among their ventral feathers, like the landing gear of a jet airplane, instead of trailing behind in the usual way, thereby economizing their energy.

These several ways of reducing the conduction of heat through the surfaces of their bodies might be inadequate to keep small birds alive through frigid nights if they did not take measures to diminish loss by convection and radiation. They do well to find overhead shelters because radiation drains much heat from bodies exposed to clear, cold nocturnal skies. Convection by chilling wind sucks even more heat from warm bodies. To avoid these losses, birds seek for sleeping the warmest, most protected spots that they can find, as in sheltered vales and the lee sides of buildings and walls. If they roost in trees, they prefer those with the densest evergreen foliage, in northern lands, especially conifers. Often they snuggle down in an abandoned open nest, of their own or some other species, in a sheltered site. They are attracted to houses, where they may sleep on porches, under eaves, amid ivy that thickly covers old walls, even inside if the building is unoccupied or a friendly householder admits them. In cities they crowd upon ledges and windowsills of tall buildings, where they are sheltered from wind, or they roost in trees that line the streets. Since the wind's velocity is greatly diminished at the level of the ground, birds of open places often sleep upon it, when possible amid grass or other herbage that further reduces the force of a gale. They may dig a little hollow in the soil and crouch down in it. Or they may dive or burrow beneath snow, where they find complete protection from wind in a nook warmer than the outer air, while their fluffed-out plumage reduces conduction and radiation to the cold white walls around them.

One might suppose that at roosts with many thousands or even millions of birds, the multitude of warm bodies close together would substantially raise the temperature of the air that envelops them. Yoram Yom-Tov and his associates investigated this question at winter roosts of the European Starling in Israel, with negative results, even where millions of birds rested among reeds at a maximum density of 405 individuals per cubic yard (530 per cubic meter). After the starlings were driven away, the temperature of the air at the roost sites did not fall. However, the birds had chosen for their nightly repose spots where the air was from 9 to 15° F. (5–8.5° C.) warmer than at nearby places. These authors computed that at a temperature difference of 15° F. with an ambient temperature in the range of 32 to 50° F. (0–10° C.), a starling weighing about 3 ounces (80 grams) saved up to 4,600 calories each night by roosting in these warmer spots. This is the energy that a starling would expend in a flight of 19 miles (31 kilometers). Many of the starlings at these roosts came from distances greater than this, spending more calories on their flight than they saved at their destination. Apparently, they had come so far for advantages other than economy of energy, such as protection from predators, or to enjoy the benefits of an information center (pages 55–56).

Although, unexpectedly, these starlings apparently gained no thermal benefit by roosting in the midst of many thousands but not in contact with any of them, the situation is different when small birds sleep pressed together. The preceding chapter told how Bushtits, who on milder nights roost with an interval between themselves and their neighbors, snuggle against each other when the temperature falls low. In the laboratory, Susan Chaplin determined that when two of these diminutive birds, each weighing only 5.5 grams, huddled together at night, at a temperature of 68° F. (20° C.), each spent only 79 percent of the energy needed by a solitary bird in the same conditions. Moreover, the paired Bushtits, having a smaller energy deficit at daybreak, foraged less actively and spent less energy during the day. When a number of Bushtits roost in a compact row, the savings for all but the two end members should be about twice as great. Closer clumping, as when many swallows or nuthatches crowd into a cavity in a tree, saves much more energy, but at the risks of suffocation or burial beneath dead companions, as was earlier mentioned.

The greatest savings of energy are gained by birds who sleep in snug cavities well protected from wind and radiation into the sky. The amount saved will depend upon the nature of the cavity. A nook beneath a metal roof would be a poor shelter on a frigid night, for the roof itself would become very cold, and the metal, a good conductor of heat, would drain away much energy from a bird resting in contact with it. Wood is a poorer conductor than metal, and dry, decaying wood is poorer than living, sappy wood. Accordingly, a bird does well to choose a cavity in a dry, rotting trunk or limb, if it can find one, and if the tree is not so far advanced in decay that it is likely to fall on a stormy night. The smaller the cavity, the less heat slight movements of the contained air will remove from the sleeper by convection, caused chiefly by the bird's warm breath. Nevertheless, the bird should not fit so tightly into its niche that it cannot fluff out its feathers, thereby reducing their conductivity. When it emerges in the morning, its bent tail may reveal that it has slept in a small space, but this will not matter if it was able to expand the plumage of its body. When a few birds occupy the hole, the saving will be greater, but they should not be too densely packed.

In winter, House Sparrows sleep singly, less often in pairs, rarely in trios, in bird boxes and other cavities, often in their old nests. In a house that, long ago in Maryland, I built for Purple Martins, a male sparrow and a female slept in separate rooms. Charles Kendeigh measured the temperature inside and outside a wooden box made of ¾-inch (2-centimeter) pine boards, the interior of which was 6 inches from front to back, 4 inches wide, and 5 inches high (15 by 10 by 12.7 centimeters). The doorway was an inch and a half (3.8 centimeters) in diameter. The box was half filled with old nesting material, including many chicken feathers. He found that at 76.2° F. (24.7° C.) the sleeping sparrow would not lose enough heat to raise the temperature of the box, and that the more the outside temperature fell below this point, the greater the difference between the inside and the outside became; for example:

Outside air temperature	62.6° F. (17° C.)	17.6° F. (−8° C.)
Difference	2.7° F. (1.5° C.)	11.2° F. (6.2° C.)
Inside box temperature	65.3° F. (18.5° C.)	28.8° F. (−1.8° C.)

At an ambient temperature of 76.2° F. (24.7° C.) the sparrow would save no energy by sleeping in the box, but at −22° F. (−30° C.), a temperature sometimes experienced by House Sparrows, it would conserve 13.4 percent of its maintenance energy—a saving that might preserve a bird's life in bleak winter weather when food is scarce. These calculations help us to understand why some birds, like Great Tits, roost amid foliage on warm summer nights but sleep in cavities in winter.

Cavities in trees, burrows in the ground, and enclosed nests built by the birds themselves of diverse materials may, according to their situation and construction, afford more or less protection from extremes of temperature than wooden boxes, but this question has rarely been investigated. An exception is the huge nest of the Sociable Weaver described in Chapter 10. About the time that Gordon Maclean made his classic study of this unique bird, a team from the University of California at Los Angeles, consisting of George A. Bartholomew, Thomas R. Howell, and Fred N. White, arrived in the Kalahari Desert in a microbus loaded with sensitive instruments to

Nest of Sociable Weavers in Kalahari Desert

study the temperature of the nests. The nest mass chosen for investigation contained seventy chambers, occupied by about 150 weavers.

During the dry austral winter, when cold southerly winds sweep across the desert and the temperature often falls well below the freezing point, four or five weavers crowd into a single room, leaving many of the chambers unoccupied. In these empty compartments the air remains through the night several degrees warmer than that outside. In occupied rooms, warmed by the bodies of the resting birds, the temperature increases directly with the number of occupants, and is generally from 32 to 41° F. (18–23° C.) higher than the surrounding air. Through most of the night, it approximates the Sociable Weavers' zone of thermal neutrality—the range of temperatures within which the basal, or maintenance, metabolism of a resting bird suffices to keep its body normally warm, without any extra expenditure of energy to generate heat or, at the upper limit of the zone, to prevent overheating, as by evaporative cooling.

However, the temperature within these thick-walled chambers, beneath the massive roof that covers all of them, does not fall at a uniform rate as through the night the whole nest mass cools, but at intervals is abruptly elevated when the birds awake and call much, by their physical activity producing more heat and keeping the rooms warmer. By sleeping in their nests, the weavers save about 40 percent of the energy they would spend if they roosted in the open during a cold Kalahari winter.

Sociable Weavers tend to make their nest masses as large as the supporting tree, telephone pole, or water tower will bear. When the nest reaches this limit, members of a colony often start new nests nearby. Most of the birds engaged in the construction of these satellite nests return to the mother colony to sleep. In a new nest with only four compartments, the temperature of unoccupied chambers differed little from that of the outside air by day or by night. When one bird slept in a chamber, the air in it remained about 4.5° F. (2.5° C.) above that of the surrounding atmosphere. The chamber occupied by two birds became 9° F. (5° C.) warmer than the ambient air and preserved almost this difference through the night. These observations demonstrate the superiority of the more massive structure in keeping the occupants warm on a cold night. They also suggest that the small dormitory nests in which many tropical birds sleep singly provide little thermal insulation but must have other advantages, such as shelter from rain, protection from predators, or simply increased comfort.

In the hot, dry summers of the Kalahari, when the minimum night temperature was around 60° F. (16° C.), from one to three Sociable Weavers (average 1.82) slept in a chamber instead of the four or five that shared a bedroom in winter. Now nearly all the rooms were occupied at night. During these warmer nights, the reduced number of occupants sufficed to keep the rooms well within the birds' zone of thermal neutrality, although empty chambers fell below the critical temperature. During the middle of the day, when the weavers seek relief from the excessive desert heat by taking a siesta in their nest chambers, they are cooler than the outside air. By day as well as by night, the massive nests preserve favorable temperatures for their industrious builders.

Only in an arid region could birds build such enormous nests and maintain them through many years; in the humid tropics they would hardly last through a long rainy season. By building these avian apartment houses and keeping them in good repair, Sociable Weavers transmit to remote descendants a valuable patrimony, as few birds do. The energy that the weavers save by sleeping in their nests reduces their individual consumption of the rarely abundant foods of the desert, thereby permitting a larger population to thrive. In a quite different way, the big nests help the weavers to maintain flourishing populations, for in them the birds can reproduce at any season, winter or summer, when the unpredictable rains enable plants to grow and set seed and insects to multiply. The Sociable Weavers' chief enemy is the Cape Cobra, which in a few days can devour all the eggs and nestlings in a nest mass with many chambers. By breeding in large, well-insulated nests in winter, when days are short and nights cold and cobras lie dormant in their dens, the weavers avoid this heavy drain upon their reproduction. By breeding in summer, in small nests on metal telephone poles that snakes cannot climb, they also foil the predator. In a manner peculiarly their own, Sociable Weavers have solved the problems of survival in an inhospitable desert.

By sleeping alone in a small nest, another desert-dweller, the tiny Verdin of the southwestern United States, saves 50 percent of the energy that it would lose if it roosted in the open on winter nights.

Dark-rumped Petrels breed in deep burrows in the crater walls of Haleakala Volcano on the Hawaiian island of Maui, at elevations between 8,000 and 10,000 feet (2,500 and 3,000 meters) in one of the world's highest colonies of nesting seabirds. Male and female parents sit alternately on the single white egg in sessions that continue for eight to twenty-three days. By illuminating a burrow with infrared light and using closed-circuit television for observation, Theodore Simons learned that incubating petrels spent about 95 percent of their time sleeping with their bills tucked into the feathers of their shoulders, 3 percent of their time resting quietly on their nest, and 2 percent arranging its materials or preening. When they slept, the petrels' respiratory rate fell from twenty-four to twelve breaths per minute. They substantially reduced the expenditure of their reserves of energy while they fasted for many days together in a cold montane climate.

Paradoxically, the best way for a homeothermal, or warm-blooded, animal to conserve energy is to abandon the effort to retain it, to permit the heat to flow freely from the body and to become, temporarily, a poikilothermal, or cold-blooded, organism. This method is practiced by a number of very small birds, whose heat-dissipating surfaces are very large in relation to their heat-generating bodies. Such birds have a high metabolism, to sustain which they must each day consume food that weighs about as much as they do. Best known and most studied of the birds that conserve energy in this manner are the hummingbirds, many of which weigh a quarter of an ounce (7 grams) or much less. On cool nights they permit their temperature to drop to that of the air around them, thereby retarding all their vital functions. Like frogs and lizards, the colder they become, the less energy they expend. In effect, they hibernate for brief periods. In *The Life of the Hummingbird,* I coined the word "noctivation" for such short-term hibernation.

Whether a hummingbird noctivates depends upon many factors, including the ambient temperature, its diet, its health, its occupation (whether breeding or not), and perhaps also its emotional state. In Augusto Ruschi's large aviary at Santa Teresa, Brazil, some hummingbirds noctivated almost every night, others rarely more than twice a week. Of two hummingbirds of the same kind resting close together, one might become torpid and the other not. However, the chief controlling factor is temperature. Like other birds, hummingbirds have warmer bodies than mammals. The temperature of a number of species, in the range of about 102 to 108° F. (39–42° C.), fluctuates a few degrees, according to whether they are active or resting quietly. In normal sleep, a hummingbird's temperature may drop from 4 to 8° F. (2.2–4.4° C.). In nocturnal torpor it falls much lower than this, almost to that of the air around it. When the night air becomes more than about 12° F. cooler than the bird's normal daytime temperature, or below about 93° F. (34° C.) the bird may noctivate.

The energy that a hummingbird saves by becoming torpid depends upon the external temperature. When the air is at about 60° F. (15.6° C.) the noctivating bird may burn only one-fiftieth to one-sixtieth of the fuel that it would need to maintain its body at the usual high temperature. Although the colder the air becomes, beyond the zone of thermal neutrality, the more energy a normally homeothermal animal will expend trying to stay warm, the reverse is true of a hibernating or noctivating creature: the colder the surrounding air, the more its metabolism is retarded, with corresponding reduction of expenditure. With rising temperature, the two rates become equal; at about 95° F. (35° C.) a hummingbird uses about the same amount of energy whether it is torpid or not, which is probably the reason why it does not noctivate unless the night air falls below this point. While incubating eggs or brooding nestlings, female hummingbirds do not become torpid at night, nor do older nestlings when left exposed. To permit their temperature to fall low would greatly retard the development of embryos and nestlings.

The bird whose body fluids congeal never awakes. Although at low altitudes in the tropics some hummingbirds roost, as they nest, with utter disregard of shelter, those of the high Andes, where nocturnal temperatures often fall well below the freezing point, seek protection in caves, mine shafts, crevices in cliffs, or buildings, where they sleep, often in company with a variety of other birds, in somewhat warmer surroundings. Moreover, when the temperature drops dangerously low, hummingbirds no longer permit their body's temperature to fall with that of the environment but raise it by metabolic processes—they are said to thermoregulate. The point at which the torpid bird's consumption of oxygen (a measure of its expenditure of energy) rises rather than falls as the air grows colder, varies with the temperature of the habitat, being lower where the average minimum night temperature is lower. In the Andean Hillstar, this reversal occurs at about 41° F. (5° C.). In hummingbirds of warmer regions, it is higher.

A noctivating hummingbird cannot fly. If lifted from its perch, it moves its limbs feebly, chirps weakly, and cannot regain its grip if replaced. The duration of torpor is variable. Some hummingbirds in Ruschi's aviary in southern Brazil noctivated for eight to fourteen hours, not becoming active until the rising sun had mitigated the early morning chill. Certain North American hummingbirds and Andean Hillstars emerge from torpor before dawn or, in captivity, before lights are turned on, in response to an internal rhythm.

Some of Ruschi's hummingbirds could be held in hand for more than an hour before they became active. Other hummingbirds regain the use of their limbs more swiftly. Aroused in the night, torpid Anna's and Allen's hummingbirds could fly more or less competently after ten or fifteen minutes. Their body temperature rose rapidly, at the rate of a degree or two per minute. Their respiration and heartbeats increased correspondingly. Not until they warm up to 86° F. (30° C.) or more can hummingbirds fly.

One reason why fairly large birds do not noctivate may be the cost in time and energy of raising their chilled bodies to daytime temperature. While torpid, they would be powerless to flutter away if an approaching predator shook their perch or a neighbor shouted warning. Probably it is because noctivation is not without increased risk of predation that this method of conserving energy is practiced only by small birds that have greatest difficulty maintaining their normally high temperature through chilly nights.

The sunbirds of the Old World, principally Africa, are in several ways the ecological counterparts of the New World's hummingbirds. They are small, slender-billed nectar-drinkers and insect-catchers, and the males of many species are brilliantly iridescent; but they cannot hover at flowers as hummingbirds do. As in the hummingbird family, some species live at great heights. The nocturnal temperatures of several of these high-altitude sunbirds were measured on Mount Kameligon in equatorial Africa, at an altitude of 11,000 feet (3,350 meters) by R. A. Cheke. On cold nights the temperatures of the Tacazze Sunbird, the Malachite Sunbird, and the Eastern Double-collared Sunbird were from 9 to 30.5° F. (5–17° C.) lower than their daytime temperatures, with a corresponding decrease in metabolism and expenditure of energy. The sunbirds' temperature often began to rise long before dawn, while the air around them was still becoming colder, suggesting an internally regulated rhythm, as in hummingbirds. Other small birds that save energy by reducing their body temperature on cold nights, a condition known as hypothermia, include Barn Swallows, Bank Swallows, House Martins, Black-capped Chickadees, Willow Tits, and Siberian Tits, but apparently none of them noctivates in the thorough manner of hummingbirds.

Among birds of low altitudes in the tropics, where frost never forms, tiny, largely frugivorous Red-capped Manakins and Golden-collared Manakins reduce their nocturnal body temperature as low as 86° F. (30° C.) when on short rations. They do not become torpid, and before dawn they spontaneously warm up to their normal daytime level. By halving the temperature gradient between themselves and a tropical night when the thermometer falls to around 68° F. (20° C.), they may diminish their expenditure of energy by about 50 percent, a saving that may be of vital importance to them during periods of relative scarcity, such as occur in some tropical regions toward the end of the wet season and in the ensuing dry season.

White-throated Swifts of western North America breed as far

north as southern British Columbia and winter from west-central California southward. Here, on extremely cold days, the swifts disappear from the air but have been found in a torpid state in the more accessible of the clefts and crevices of the rocky cliffs where they nest and sleep. Since when cold and undernourished they remain dormant even in the daytime, they can hardly be said to noctivate, as hummingbirds do. In the laboratory, swifts who had lost much weight became torpid much more readily than heavier individuals. At air temperatures as low as 39° F. (4° C.) chilled swifts could become active again if their body temperature did not fall below 68° F. (20° C.), their temperature during arousal rising at the rate of about 0.7° F. (0.4° C.) per minute. At a body temperature as low as 95° F. (35° C.) they appeared to be fully active. Recovery from torpidity has been reported of the Chimney Swift and a few others. The ability of undernourished nestling Common Swifts to reduce their temperature during the night and become normally active, able to take their meals, the following day, helps them to survive periods of stormy weather when their parents cannot catch enough flying insects for them.

Hibernation might be regarded as prolonged sleep—noctivation, as in hummingbirds, prolonged for months. From ancient times, Europeans believed that some of the birds that disappeared in winter passed this season in a state of torpor. Surprisingly, Aristotle, who was not ignorant of avian migration, included among species that hibernate even some that migrate as conspicuously as swallows. As late as the eighteenth century, when much more was known about bird migration, some of the best naturalists in Europe seriously considered the possibility that swallows and swifts hibernate through the winter. Swallows were believed to dive beneath the water of lakes and ponds, to sleep away the cold months in the mud at the bottom. The occasional discovery of a torpid swift or swallow on cold days probably lent weight to this myth. With the expansion of knowledge of birds, ornithologists became skeptical of all reports of hibernation by feathered creatures. Not until 1948 and 1949, when Edmund C. Jaeger published his observations on profound torpor in a Common Poorwill, did hibernation by a bird gain full credence by modern naturalists.

A poorwill, a small, short-tailed nightjar named for its call, was found in the winter of 1946–1947, in the Chuckwalla Mountains of the Colorado Desert in California, in a state of profound torpor. In a depression in the face of a granite rock barely large enough to contain it, the bird rested with head inward and tail exposed. Taken in hand, it gave hardly any sign of life. In the following winter, presum-

Common Poorwill hibernating

ably the same individual was found sleeping profoundly in the same niche in the rock from November 26, 1947, until February 14, 1948, an interval of about eighty-five days. By February 22 it was sufficiently awake to fly away when removed from its crypt. Banded during the second winter, the same bird returned to its hibernaculum for the third winter in November 1948, but vanished before the year's end, probably having been carried away by some predator or inquisitive person.

The hibernating poorwill's temperature remained in the range of 64.4 to 67.6° F. (18 to 19.8° C.) while air temperatures late in the mornings fluctuated from 63.5 to 75.4° F. (17.5 to 24.1° C.). No sound of heartbeat could be heard with a medical stethoscope. No moisture was deposited on a cold metal mirror held in front of the bird's nostrils; no movement of breathing could be detected. During the entire period of observation, it voided no wastes. In six weeks, beginning in early January, its weight diminished by about 1 gram, from 45.61 to 44.56 grams. Soon after it began to hibernate in the following winter, it weighed 52.68 grams. On successive visits, only slight shifts in the position of the bird's head were noticed. By reducing its metabolism to a minimum, the insectivorous poorwill could remain in familiar territory and sleep away the coldest months when insects were fewest, which was a course probably less hazardous than migration.

In contrast to small birds, for whom the conservation of heat in cold weather is a major problem, certain large birds well adapted to frigid climates have such low critical temperatures that they hardly need to take special measures to conserve energy. Even at −22° F. (−30° C.) Glaucous Gulls above the Arctic Circle do not need to exceed their basal metabolism to keep normally warm. Small birds withstand exposure to Arctic cold better than mammals of comparable size. Black-capped Chickadees, Hoary Redpolls, and other birds are abroad and active on the coldest days, while lemmings, ground squirrels, and mice remain much of the time in their burrows and runways beneath sheltering snow, emerging only briefly.

References: Bartholomew 1985; Bartholomew, Howell, and Cade 1957; Bartholomew, White, and Howell 1976; Beal 1978; Brodkorb 1955; Buttemer et al. 1987; Carpenter 1976; Chaplin 1982; Cheke 1971; Irving 1960; Jaeger 1948; 1949; Kendeigh 1961; Moore 1945; Simons 1985; Skutch 1973; Staebler 1941; Walkinshaw 1953; Wetmore 1936; White, Bartholomew, and Howell 1975; L. Williams 1947; Yom-Tov, Imber, and Otterman 1977.

THIRTEEN

How Long Do Birds Rest?

I started to head this chapter "How Long Do Birds Sleep?" Then it occurred to me that I do not know how long I sleep, and even less do I know how long birds sleep. To be sure, I know when I go to bed and when I arise, but often I lie for a while before I slumber, or I wake in the night; I do not look at the clock every time I wake, and I cannot consult it when I fall asleep. Similarly, birds may settle in their roosts or enter their dormitories many minutes before they drowse; and they, too, are often wakeful at night, as attested by their voices or movements, for intervals that it is hardly possible to measure. The best we can do is to time their arrival at, and departure from, their places of rest, and even this is not always easy. The principal period of rest of diurnal birds is, of course, the night; but often they retire while much daylight remains, or linger in their roost or dormitory long after daybreak. Moreover, with the possible exception of small birds who must make the best of every minute of a short winter day to glean enough food to tide them over a long, cold night, most birds appear to rest at intervals through the day, sometimes taking a siesta in the warmest hours.

The hour at which birds go to rest in the evening and become active in the morning and, accordingly, the duration of their nocturnal rest, varies with the species, the sex, the individual, the latitude, the season, the weather, and the activities in which the birds are engaged, especially nesting. With so many variables, it would seem that we must welter in a choppy sea of details. However, some illuminating generalizations may be drawn.

Many studies in diverse regions reveal that birds become active in the morning when the light is dimmer than it will be when they retire in the evening, probably because at dawn they are rested and hungry, while at day's end they are well fed and weary. They seem to follow the old adage "Early to bed and early to rise . . ." Unlike most people, especially city-dwellers, they rest longer before than after astronomical or solar midnight, which may differ from local time, especially daylight-saving time. However, when we compare species in the same locality, those that go soonest to bed are not always the earliest risers, but rather the reverse, as is especially evident among dormitory-users. Woodcreepers and Blue-throated Green Motmots are among the last birds to seek their dormitories in the evening and the first to emerge in the morning, entering and leaving when they are difficult to see in the dim light. At the other extreme, woodpeckers and some tropical swallows retire early and arise late, but they are often unpredictable. Intermediate in times of retiring and arising are araçaris, wrens, and Bananaquits.

Birds who retire early and become active late tend to be more variable in their times of going to bed and arising than those who go to rest late and begin their day early. In California, Laidlaw Williams found that Chestnut-backed Chickadees, who nearly always entered their nooks before sunset, were less regular in their hour of retiring, and did so at a greater range of light intensities, than their neighbor Bewick's Wren, who did not retire until sunset or later. Birds seek their roosts or dormitories earlier on rainy or darkly cloudy than on clear evenings, and they become active earlier on fair than on wet mornings. Not only are the movements of the early-to-bed birds more responsive to variations in such environmental conditions as temperature, precipitation, and wind; they also appear to be more subject to individual whims. The irregularity in times of morning departure is not necessarily caused by earlier or later waking, for the birds may gaze through their doorways for many minutes before they depart. Similarly, earlier entry into a dormitory does not signify that a bird will fall asleep sooner than a neighbor who retires an hour later.

Among passerine birds, males consistently retire later and arise earlier than females, even at times when neither attends a nest. When a female Southern House-Wren hears the joyous song that through much of the year greets the new day as soon as her mate flies from the nook where he sleeps, she may leave her neighboring dormitory almost immediately or linger a few minutes longer, rarely as much as ten. In Costa Rica, the male Bananaquit retires somewhat later than his mate, as did a male Chestnut-backed Chickadee in California. In winter and spring in England, the male of a pair of European Starlings entered the hole in which both slept after his mate and left earlier than she did. The males of Song Sparrows, Black Phoebes, and many other birds take a little less rest than their mates.

Among woodpeckers, the order of retiring and arising tends to be reversed. The male of a pair of Red-crowned Woodpeckers regularly entered his hole earlier in the evening than his mate, and delayed there later in the morning. Sometimes the difference in their times of entering or leaving was from twenty minutes to half an hour. My fourteen records contain a single exception to this rule, a morning when the male left his hole about one minute before the female quitted hers. The industrious male of this pair always enjoyed the newer and sounder dormitory, while his less careful partner was often satisfied with the more or less dilapidated shelters that he had abandoned.

Male and female Golden-naped Woodpeckers, sleeping in the same hole, have no fixed order of entering or departing; sometimes the male enters or leaves first, sometimes the female; and after their sons, who continue to lodge with them, are so well grown that they can no longer be distinguished from their father, the mother may enter or leave between two of the males. Nevertheless, I have records of seventeen times that a male entered first for the night, to five times by the female; and of sixteen times that the female emerged first in the morning, to nine times by a male. These observations were made while the Golden-napes were neither incubating nor brooding nestlings. During incubation, the male usually leaves first. At times the female emerges first from the hole and clings to the trunk near the doorway until her mate flies away, then promptly re-enters to cover the eggs.

The observations on passerines and woodpeckers are not so irreconcilable as at first appears. Unlike passerines, the male woodpecker commonly incubates through the night while, in most species, his mate sleeps in another hole. In both cases, the sex that incubates by night takes the longer rest, even at seasons when breeding is in abeyance.

The major differences in the length of the night rest are caused by differences in the length of the night, which varies with latitude and, everywhere except at the equator, with the changing seasons. Observations on the duration of resting on or near the equator, where twilight is relatively brief, are unavailable; but for diurnal birds it is probably 12 hours, or a little less, throughout the year. In Costa Rica and Panama, between 9 and 10 degrees north latitude, dormitory-users, for whom accurate records are most readily obtained, rest for about 11½ to 12½ hours, with variations appearing to depend more upon the species and activities of the birds, and upon weather, than upon the position of the Sun in the sky.

In the North Temperate Zone, many have recorded the hour when

Golden-naped Woodpeckers at nest hole

birds begin to sing at dawn, which may be before or after they leave their roosts, while others have made records of the times when the birds enter or leave their roosts or other sleeping places. The data are frequently reported in terms of light intensity or civil twilight (which begins or ends when the rising or setting sun is six degrees below the horizon), making it difficult to ascertain just how long the birds spend at their roosts, particularly when the times of both retiring and arising are not given. Fortunately, a few observers have supplied all the necessary information. At Carmel, California, 36 degrees north latitude, Laidlaw Williams watched a Bewick's Wren whose night rest lasted about 12½ hours in February, 11 hours in March, and 10 hours in August. In the same locality, a Chestnut-backed Chickadee rested 13 hours and 41 minutes in February, 12 hours in March, and 10½ hours in August. Near Knoxville, Tennessee, also at 36 degrees north, T. David Pitts found Carolina Chickadees resting in their holes for about 15 hours when winter days were shortest. At Champaign, Illinois, 40 degrees north, the House Sparrow who slept in the box where Charles Kendeigh recorded temperatures occupied this shelter for 15 hours, within a few minutes, in December and January. At Bournemouth, on the south coast of England at nearly 51 degrees north, a female Green Woodpecker, as reported by F. C. R. Jourdain, occupied her nest box for approximately 16 hours on December nights. At Gladenbach in Germany, at the same latitude, a Great Spotted Woodpecker rested for about 15½ hours in December and 7½ hours in June.

At Cambridge, England, 52 degrees north, Edward A. Armstrong recorded for a year the times of retiring and arising of Winter Wrens, mostly his banded male, Silver. The graph that he compiled shows that Silver rested slightly more than twice as long in December as in June, about 15½ hours around the winter solstice but only 7½ hours when nights were shortest. Through much of the year, the wren retired around sunset, but in December's short days he might remain active, seeking food, for about half an hour after sunset. In the long days of June, he often went to rest about an hour before sunset. In the morning, he usually became active well before sunrise; in midsummer, he arose around three o'clock in the morning, nearly an hour before the sun appeared on the horizon. His time of arousal conformed more closely to civil twilight than to sunrise.

At Churchill, Manitoba, on Hudson Bay at 58 degrees north, in June and July, when daylight is almost full from around 2:00 A.M. until 10:00 P.M., A. M. Baumgartner found American Tree Sparrows active for 17¾ hours, leaving 6¼ hours for rest. In the continuous daylight above the Arctic Circle in midsummer, most birds appear to need some rest. Different species take it at different times, mostly around midnight, when the Sun is lowest above the horizon. Various authors report Snow Buntings as resting from May onward for about three to five hours, more before than after midnight. In Swedish Lapland at 68 degrees north in midsummer, Armstrong learned that Willow Tits feeding nestlings rested for seven or eight hours, from around 6:00 or 7:00 P.M. to 2:00 A.M. Fieldfares attending young were least active for several hours before midnight. Parent Black-bellied Dippers rested from about 3:00 to 8:00 P.M. A few hours before midnight Armstrong noticed a lull in the singing of passerine birds, which immediately after midnight burst into full chorus. None of these observations reveal how long any individual bird rested.

Watching a nesting cliff of Black-legged Kittiwakes on the Island of Jan Mayen at 71 degrees north in the Arctic Ocean in late July, J. M. Cullen found that individual adults slept, with bills resting on their backs, for about seven hours each day, about as frequently at noon as at midnight.

In addition to their long nocturnal repose at lower latitudes where moonless nights are dark, birds frequently rest in the daytime. In early spring, a female Northern Mockingbird flew into a privet shrub beside Amelia Laskey's house in Nashville, Tennessee, tucked her head into her feathers, and napped in mid-morning. Years ago, I occupied a house propped upon high concrete pillars in a new clearing in the rain forest near the Golfo Dulce in Costa Rica. Returning to my dwelling late on an October morning, I was surprised to find a diminutive female Blue-crowned Manakin perching upon a white handkerchief hung over a wire to dry, almost beneath the center of the house. To find a forest bird in this domestic setting was so unexpected that I sat nearby to watch her. She was not asleep but only resting. After five minutes, she darted back into the forest, leaving droppings on the concrete floor beneath her perch to attest that her rest had been long.

In torrid regions where the midday temperature may exceed that of birds, exposing them to heat stress, many seek relief in the coolest, shadiest spots they can find. Activity in the feathered world is at its lowest ebb. Sociable Weavers return from foraging over the Kalahari Desert to take a long siesta in their nest chambers, which at this time of day are often cooler than the air outside, as told in Chapters 10 and 11. Red-billed Queleas roost in their thousands or millions in sheltering shrubs and small trees in the middle of the day as well as by night, to forage again as the Sun declines toward the west and the air cools. In a very different climate, female Scarlet-rumped

Tanagers nesting around our house in Costa Rica take their longest diurnal rests around noon, sometimes sitting quietly in their nests, if not sleeping, for over one hundred minutes, which is very much longer than they incubate continuously at other times of day.

Although birds commonly continue to forage in gentle showers, a heavy downpour makes them seek shelter. Incubating and brooding birds who come and go from their nests many times each day sit for long intervals while rain falls, or brood feathered nestlings that they have ceased to cover on fair days. Other birds disappear amid the foliage, which in the tropics contains many leaves ample enough to form a roof above a small bird—or a man. Those with dormitories sometimes retire into them, and the time they spend sheltering there might, without exaggeration, be included among their hours of rest.

While I dwelt in a narrow Costa Rican valley at an altitude of about 3,000 feet (900 meters), I followed the activities of a number of birds that slept in holes or covered nests that they built for themselves. At three o'clock on an afternoon in late May, when a long-threatened rain began to fall and soon increased to a heavy shower, I put on my raincoat for a round of visits. The male Bananaquit who slept in the guava tree behind my cabin was not at home. The Southern House-Wrens' gourd in the orange tree was unoccupied. The female Plain Wren was not in her dormitory nest. Climbing up to the clearing at the edge of the forest, I found the male Red-crowned Woodpecker looking out of his hole newly carved in the soft wood of a Burío tree. He was not then nesting but preparing for a second brood. I could not find his mate, whose absence from all the holes where she sometimes slept led to the conclusion that she was still out in the rain. The Fiery-billed Araçaris had not entered the old hole of a pair of Pale-billed Woodpeckers where they lodged.

Last of all, I visited the Golden-naped Woodpeckers, whose two surviving fledglings had been flying about for a fortnight. Their mother was in the lofty hole in a massive, fire-charred trunk, gazing out at the wet landscape. Soon a male member of her family flew up and she made way for him to enter, without relinquishing her post in the doorway. Before long the Sun, falling westward, began to glare through the thinning clouds, while the rain fell with undiminished intensity. The mother, becoming restless, flew out into the downpour; the young male promptly followed. His departure left the hole vacant; the father and the other son had remained out in the rain. A few days earlier, I had seen the father take shelter from an afternoon rain, while the mother stayed outside. On another afternoon, both parents kept dry in the hole with their fledglings.

The response to rain of other Golden-naped Woodpecker families has been equally unpredictable. Sometimes they enter their holes but at other times they stay out in the wet, often clinging upright to the lower side of a leaning trunk. Parents may keep dry in their hole while their fledglings remain outside. If a hard rain continues well past the middle of the afternoon, the whole family may retire into their dormitory as early as four o'clock and stay into the night, the fledglings going hungry to bed. In addition to the Golden-naped Woodpecker and the Red-crowned Woodpecker, the Acorn Woodpecker, Downy Woodpecker, and Northern Flicker have been found in their cavities on wet days. Apparently, the habit of entering their holes when it rains is widespread among woodpeckers.

On the afternoon following my first round of visits in a raincoat, I repeated them. The rain started at half-past two with a torrential downpour that spent too much water to be sustained, then settled down to a heavy shower that continued through much of the afternoon. Not a single dormitory was occupied. Apparently, the intensity of rain is not the only factor that determines whether birds will enter their dormitories. Much probably depends upon how far from home they happen to be when the shower begins, whether they are hungry or satisfied, their general physiological state, and the time of day.

In October, the drenching climax of our rainy season, I often found the male Red-crowned Woodpecker in his hole between four and five o'clock in the afternoon, an hour or two before nightfall. He would remain with his head in the doorway until daylight waned, then disappear into the bottom of his chamber. Next morning he would linger within until nearly six o'clock, prolonging his period of continuous rest to between thirteen and fourteen hours. In the evenings his mate would loiter in neighboring dead trees, pecking here and there or clinging motionless to a branch, long after he had retired. She did not enter her hole until the light grew dim.

Although I wrote that woodpeckers are among the first birds to retire in the evening and the last to become active in the morning, exceptions occur. In the densely populated central valley of Costa Rica, Hoffmann's Woodpeckers, relatives of the Red-crowns, entered their holes so late in the evenings, and left so early in the mornings, that it was difficult to distinguish their sexes in the dim light. Possibly in this region, where birds lacked legal protection, they entered and left in the twilight to be less conspicuous to boys who might use them as targets for their rubber catapults.

Acorn Woodpeckers may seek their dormitories early or remain out very late. In Costa Rica, before nesting began, five lodged in a massive, branchless, fire-gnawed trunk in a hillside pasture at the forest's edge. After spectacular aerial flycatching and chasing each other with much animated calling in the twilight, they flew in a cluster to their dormitory, where all tried to squeeze at the same time through a doorway wide enough for only one. In the dim light, they entered single file. On a high Guatemalan mountain, I watched an Acorn Woodpecker retire after the earliest stars shone out. On a December morning, I again watched this same dormitory. While the light was still dim, a white forehead appeared in the orifice; a woodpecker had awakened and was looking through his doorway at the frosty dawn. While he remained peering forth, almost motionless, the bird world became active. Soon I heard the calls of other Acorn Woodpeckers who lodged in neighboring pine trees. About five minutes later, the bird in the doorway finally came out, after looking forth for over half an hour. He alighted on the stub of a broken-off branch and called loudly *rack up, rack up.* The moment he emerged, a similar head replaced his in the entrance. In a minute the first woodpecker flew back to the doorway and hung motionless just below it, while the rising Sun's first rays warmed him on this chilly morning. After ten minutes here, he bestirred himself, flew to the top of a tall dead pine tree, and called. A minute later the second woodpecker joined the first in the open. It was then a few minutes past seven o'clock, an hour after daybreak. Unlike many other birds, woodpeckers often appear not to be hungry when they arise; they may delay for many minutes in or beside their dormitories, prolonging their rest, before they seek breakfast.

While David Lack and his collaborators sat in the tower of the Museum of Science at Oxford University, studying Common Swifts that nested there, they sometimes heard raindrops pelting on the slates outside. Almost at once, several swifts would fly into their nests for shelter and stay until the rain stopped. At the experiment station at Amani in Tanganyika Territory (now Tanzania) White-rumped Swifts nested and slept in closed, mud-walled nests that Abyssinian Swallows had built beneath the shelter of the eaves, where they were studied by R. E. Moreau and his assistants. As a rule, these swifts did not enter their nests to escape rain, but during one downpour both parents stayed inside for over half an hour with their nearly fledged young.

The pair of Blue-and-white Swallows who for two years slept beneath the roof tiles of my present abode, where they twice tried without success to raise a brood, also took long periods of repose. In the long interval between nesting seasons, their return to the roof in the evening depended largely on the weather, but at latest was earlier than that of the Southern House-Wrens who also slept beneath the tiles, and before the Southern Rough-winged Swallows plunged into a neighboring field of sugarcane to roost. On clear evenings the Blue-and-white Swallows might on rare occasions remain out until half-past five, but usually, especially around the time of the winter solstice, they went to rest at five o'clock or earlier, an hour before nightfall. On the rainy afternoons so frequent in the second half of the year, they sought shelter much earlier. On some gloomy, wet afternoons, they entered their nook soon after three o'clock. If the rain abated before sunset, they might emerge to fly around and catch insects for a while, before their early and final return for the night. If the rain continued hard until evening, they would stay beneath the tiles, however hungry they might be.

At daybreak, the pair of Blue-and-white Swallows would linger in their nook until smoke began to rise from the wood fire in the neighboring kitchen. Long after nearly all their feathered neighbors had become active, only ten or fifteen minutes before the Sun flared up above the forested crest of the eastern ridge, they would fly out from beneath the tiles, where on some nights they had rested for nearly fifteen hours. Usually they emerged almost together; but on some mornings, especially as the nesting season approached, one would delay for a minute or two, rarely as much as five minutes, after its mate had flown beyond view. As soon as they became active, they nearly always flew northward toward the high, craggy summits of the Chirripó massif until I could see them no longer. Then they would be absent until mid-afternoon or evening. How far they went I could never learn.

Black-capped Swallows, grayish brown above and largely white below, with a glossy black hood and conspicuously forked, blackish tail, are resident in the mountains from the Mexican state of Chiapas to El Salvador and Honduras. In October, I found a number of them sleeping in a deep burrow, apparently dug by Blue-throated Green Motmots, at the top of a high earthen bank beside a Guatemalan highway, about 9,000 feet (2,750 meters) above sea level. They were difficult to count as they retired in the evening. Arriving as the light began to fade, they circled around and around in front of the burrow, delaying to enter with typical hirundine procrastination. Often they flew directly toward the tunnel as though intending to enter, only to veer aside at the last moment and continue to gyrate around; sometimes they flew up under the foliage that draped in front of the entrance, or struck a leaf with a resounding *plop,* then darted out again to trace more circles in the air with their comrades. At times two or three together, all rushing about in the vicinity, converged toward the burrow as though moved by a single impulse. Since they could not all enter simultaneously, they solved the difficulty by all turning away to continue their circlings. Rarely, one or two belated swallows would fly up the mountainside and shoot into the burrow without the usual procrastination. By this time it was nearly dark, and after so much confusing movement I was uncertain how many birds had entered.

After a long climb by moonlight, I arrived at dawn to count these Black-capped Swallows as they left their burrow. This was situated on an exposed shoulder of the mountain, and as, heated by my climb, I sat watching on the bank on the opposite side of the highway, the cold, brisk breeze cooled me rapidly. Around seven o'clock, I heard other Black-capped Swallows twittering down the slope behind me; but those for whom I watched did not appear. The rays of the rising Sun had descended to the foot of the high bank when, at eight o'clock, three swallows shot out of the burrow, chirping and twittering as they flew promptly beyond view. Another hour and a half dragged by before, at about half-past nine, the remaining five birds darted out in rapid succession and followed down the mountainside. They had slept, or at least enjoyed the warmth of their burrow, for more than half of that beautiful and sunny but unpleasantly cool morning, over three and a half hours after most other birds had arisen with the dawn, and a full hour and a half longer than their bedfellows.

There was no reason to suppose that these swallows were in any way stressed or undernourished, like the migratory swallows who crowded into whatever shelter they could find when overtaken by unseasonable cold. The verdant mountainside where they lodged was bright with flowers before which hummingbirds hovered. They were playfully active before they entered in the evening. A short flight down the slopes would have taken them to warmer regions where insects were probably flying more freely. Not hungry enough to bestir themselves and face the chilly but far from freezing mountain air, they were simply luxuriating. On December 4, these swallows retired between 5:43 and 5:51 P.M. Next morning four flew out at 8:20 A.M. and three more at 8:27, after having rested for 14½ hours, an interval much longer than the night, even in December, at 15 degrees north latitude.

I have not known an adult Southern House-Wren to take shelter from a daytime rain; on the wettest evenings they retire only a little earlier than on fair ones. Young wrens occasionally enter their dormitory to escape a drenching. A shower that began at a quarter past five on a gloomy afternoon sent three month-old wrens into the gourd where they slept. As soon as the rain ceased, they came out again, and did not retire for the night until considerably later. Two days afterward, a shower beginning at five o'clock sent one of the three hurrying into the gourd. Since neither of its siblings followed, it promptly came out, preferring a wetting in company to dryness in solitude.

It was amusing to watch Banded-backed Wrens leave their large covered nests high in the Guatemalan mountains on wet or blustery mornings. On the cold and misty morning of March 1, I arrived in view of a dormitory occupied by six wrens at 6:20, when birds who lacked such comfortable bedrooms were already active. Nearly half an hour passed before one of the wrens appeared in the doorway. One glimpse of the cold, wet, gray morning was enough to convince the bird that it was too early to arise; promptly it withdrew into the nest, where all remained out of sight for twenty minutes more. Then a wren advanced into the doorway, where for many minutes it stood with only its head and spotted breast visible, gazing upon the driving mist and dripping foliage. At last, probably feeling hungry, it crept outside, but instead of flying to the ground as usual, it perched beside the nest and waited quietly, with out-fluffed plumage, for its more somnolent bedfellows. Finally, at 7:15, two wrens perched beside the first, while two more flew directly to the ground. One even ventured to sing a harsh refrain, in keeping with the mood of that harsh day. The sixth and last did not emerge until after its companions had started to break their fast among the wet bushes, where it promptly joined them.

One morning in September, when the mountain was darkly enveloped in a dense cloud, I watched Banded-backed Wrens arise at another dormitory. While the light was still dim, one of them stepped forth from the nest, paused on a branch in front of the doorway, turned around, and sang loudly at its resting comrades. Thereupon, a second wren, apparently annoyed by this loud singing when there was hardly enough light to begin the day, came out and bumped directly into the singing bird. The two wrens grappled, bill clasping bill, and fluttered downward, sparring, until I lost sight of them among the bushes—the only fight I ever witnessed among these birds. Relieved of the disturber, the other seven wrens emerged slowly, one by one, the last a quarter of an hour after the earliest riser had shattered the quiet of the dormitory. As often happened on

Scissor-tailed Flycatchers at sunset

such inclement mornings, some of the wrens turned around and re-entered the nest for brief intervals.

Some birds choose roosting sites that permit them to prolong their active day. Each summer in late July and August, at Tioga Pass in the mountains of California, Diana F. Tomback watched Clark's Nutcrackers spend the last hours of daylight on the sunlit west-facing slope of Mount Dana, where they extracted seeds from the cones of Whitebark Pines and fed clamorous fledglings. Then they perched on high treetops to call in a social gathering. As the Sun's last rays faded from the mountaintop, from thirty to fifty nutcrackers flew through Tioga Pass to roost in dense forest on the east-facing slope of Gaylor Ridge, a distance of 1.25 miles (2 kilometers). Here, next morning, they enjoyed the earliest warming rays of the rising Sun after a cold night about 10,000 feet (3,050 meters) above sea level. By passing the last hour of the day on a west-facing slope, and sleeping on an east-facing slope where day dawned sooner, the birds not only prolonged their active day at both ends in a busy season but also kept warmer in the thin mountain air while the rising or setting Sun was low.

Another bird that timed its movements to remain in the Sun's last rays, probably with a different motive, was the Scissor-tailed Flycatcher. One evening in Guanacaste Province in northwestern Costa Rica, while the Sun sank above a purple ridge toward the Pacific Ocean, we sat on a wooded hilltop, watching these wintering birds go to roost in low trees in drier parts of a broad marsh. From the surrounding country over which they had scattered by day, the flycatchers, traveling independently rather than in flocks, flew gracefully over the hilltop toward the marsh in an open stream that continued for a quarter of an hour. At first, most of the birds flew at no great height above the treetops, where the horizontal beams of the sinking Sun set their lovely orange sides aglow. As the Sun fell lower and the shadows rose, the birds also flew higher and higher, continuing to pass overhead in the sunshine that illuminated their sides, while we sat in deepening twilight. Not until the birds who flew highest were in Earth's shadow did the procession cease.

By rising into the last rays of the setting Sun before they slept, the flycatchers hardly increased the length of their day, but their action was symbolic. Diurnal birds, the great majority of the feathered kind, feel safest in the friendly daylight; in the darkness of night they are less able to escape the many predators that hunger for their flesh, and when the temperature falls they are in danger of exhausting their sources of energy in the effort to maintain their vital heat.

In the foregoing pages, we reviewed the diverse measures they take to confront these perils.

References: Armstrong 1954; 1955; Baumgartner 1937; Blume 1965; Cullen 1954; Jourdain 1936; J. R. King 1986; Lack 1956; Laskey 1943; Moreau 1942b; Nice 1943; Oberlander 1939; Pitts 1976; Skutch 1960; 1969b; Tomback 1978; L. Williams 1941; Zammuto and Franks 1981.

Bibliography

Anderson, A. H., and A. Anderson. 1957. Life history of the Cactus Wren, Part I: Winter and prenesting behavior. *Condor* 59:274–296.

———. 1962. Life history of the Cactus Wren, Part V: Fledging to independence. *Condor* 64:199–212.

Armstrong, E. A. 1940. *Birds of the grey wind.* London and New York: Oxford University Press.

———. 1954. The behaviour of birds in continuous daylight. *Ibis* 96:1–30.

———. 1955. *The Wren.* London: Collins.

Ashmole, N. P. 1962. The Black Noddy *Anous tenuirostris* on Ascension Island, Part I: General biology. *Ibis* 103b:235–273.

———. 1963. The biology of the Wideawake or Sooty Tern *Sterna fuscata* on Ascension Island. *Ibis* 103b:297–364.

Austin, G. T. 1976. Behavioral adaptations of the Verdin to the desert. *Auk* 93:245–262.

Back, G. N., M. R. Barrington, and J. K. McAdoo. 1987. Sage Grouse use snow burrows in northeastern Nevada. *Wilson Bull.* 99:488–490.

Bagg, A. M. 1943. Snow Buntings burrowing into snowdrifts. *Auk* 60:445.

Bartholomew, G. A., Jr. 1942. The fishing activities of Double-crested Cormorants on San Francisco Bay. *Condor* 44:13–21.

———. 1943. Daily movements of cormorants on San Francisco Bay. *Condor* 45:3–18.

———. 1985. Torpidity. In Campbell and Lack, eds. 1985.

Bartholomew, G. A., T. R. Howell, and T. J. Cade. 1957. Torpidity in the White-throated Swift, Anna Hummingbird, and Poor-Will. *Condor* 59:145–155.

Bartholomew, G. A., F. N. White, and T. R. Howell. 1976. The thermal significance of the nest of the Sociable Weaver *Philetairus socius:* Summer observations. *Ibis* 118:402–410.

Baumgartner, A. M. 1937. Food and feeding habits of the Tree Sparrow. *Wilson Bull.* 49:65–80.

Beal, K. G. 1978. Temperature-dependent reduction of individual distance in captive House-Sparrows. *Auk* 95:195–196.

Beebe, W., G. I. Hartley, and P. G. Howes. 1917. *Tropical wild life in British Guiana,* vol. 1. New York: N.Y. Zool. Soc.

Bennett, L. J. 1938. *The Blue-winged Teal: Its ecology and management.* Ames, Iowa: Collegiate Press.

Bent, A. C. 1923. Life histories of North American wildfowl, Part 1. *U.S. Natl. Mus. Bull.* 126.

———. 1925. Life histories of North American wildfowl, Part 2. *U.S. Natl. Mus. Bull.* 130.

———. 1926. Life histories of North American marsh birds. *U.S. Natl. Mus. Bull.* 135.

———. 1927. Life histories of North American shore birds, Part 1. *U.S. Natl. Mus. Bull.* 142.

———. 1929. Life histories of North American shore birds, Part 2. *U.S. Natl. Mus. Bull.* 146.

———. 1932. Life histories of North American gallinaceous birds. *U.S. Natl. Mus. Bull.* 162.

———. 1937. Life histories of North American birds of prey, Part 1. *U.S. Natl. Mus. Bull.* 167.

———. 1938. Life histories of North American birds of prey, Part 2. *U.S. Natl. Mus. Bull.* 170.

———. 1940. Life histories of North American cuckoos, goatsuckers, hummingbirds and their allies. *U.S. Natl. Mus. Bull.* 176.

———. 1942. Life histories of North American flycatchers, larks, swallows, and their allies. *U.S. Natl. Mus. Bull.* 179.

———. 1946. Life histories of North American jays, crows, and titmice. *U.S. Natl. Mus. Bull.* 191.

———. 1948. Life histories of North American nuthatches, wrens, thrashers and their allies. *U.S. Natl. Mus. Bull.* 195.

———. 1949. Life histories of North American thrushes, kinglets, and their allies. *U.S. Natl. Mus. Bull.* 196.

Bent, A. C., and collaborators. 1968. Life histories of North American cardinals, grosbeaks, buntings, towhees, finches, sparrows, and allies. 3 vols. *U.S. Natl. Mus. Bull.* 237.

Blume, D. 1965. Ergänzende Mitteilungen zu Aktivitätsbeginn und -ende bei einigen Spechtarten unter besonderer Berücksichtigung der Grauspechtes (*Picus canus*). *Vogelwelt* 86:33–42.

Boersma, P. D. 1977. An ecological and behavioral study of the Galápagos Penguin. *Living Bird* 15(for 1976):43–93.

Bourne, G. R. 1975. The Red-billed Toucan in Guyana. *Living Bird* 13(for 1974):99–126.

Brewster, W. 1949. In Bent 1949.

Brodkorb, P. 1955. Number of feathers and weights of various systems in a Bald Eagle. *Wilson Bull.* 67:142.

Brown, C. R. 1978. Post-fledging behavior of Purple Martins. *Wilson Bull.* 90:376–385.

———. 1980. Sleeping behavior of Purple Martins. *Condor* 82:170–175.

———. 1986. Cliff Swallow colonies as information centers. *Science* 234: 83–85. Abstract in J. Field Ornithol. 58:248 (1987).

Bucher, E. H. 1970. Consideraciones ecológicas sobre la paloma *Zenaida auriculata* como plaga en Córdoba. Ministerio de Economía y Hacienda, Dirección Provincial de Asuntos Agrarios, *Ser. Ciencia y Técnica,* No. 1:1–11.

Buss, I. O. 1942. A managed Cliff Swallow colony in southern Wisconsin. *Wilson Bull.* 54:153–161.

Buttemer, W. A., L. B. Astheimer, W. W. Weathers, and A. M. Hayworth. 1987. Energy savings attending winter-nest use by Verdins (*Auriparus flaviceps*). *Auk* 104:531–535.

Caccamise, D. F., L. A. Lyon, and J. Fischl. 1983. Seasonal patterns in roosting flocks of Starlings and Common Grackles. *Condor* 85:474–481.

Campbell, B., and E. Lack, eds. 1985. *A dictionary of birds.* Calton, England: T. and A. D. Poyser.

Carpenter, F. L. 1976. Ecology and evolution of an Andean hummingbird (*Oreotrochilus estella*). *Univ. Calif. Publ. Zool.* 106:1–73.

Cater, M. B. 1944. Roosting habits of martins at Tucson, Arizona. *Condor* 46:15–18.

Chaplin. S. B. 1982. The energetic significance of huddling behavior in Common Bushtits (*Psaltriparus minimus*). *Auk* 99:424–430.

Cheke, R. A. 1971. Temperature rhythms in African montane sunbirds. *Ibis* 113:500–506.

Cherrie, G. K. 1916. A contribution to the ornithology of the Orinoco region. Mus. Brooklyn Inst. Arts and Sci., *Sci. Bull.* 2:133a–374.

Christy, B. H. 1940. Mortality among Tree Swallows. *Auk* 57:404–405.

Collias, N. E., and E. C. Collias. 1964. Evolution of nest building in the weaverbirds (Ploceidae). *Univ. Calif. Publ. Zool.* 73.

———. 1978. Cooperative breeding in the White-browed Sparrow Weaver. *Auk* 95:472–484.

———. 1980. Behavior of the Grey-capped Social Weaver (*Pseudonigrita arnaudi*) in Kenya. *Auk* 97:213–226.

Combellack, C. R. B. 1954. A nesting of Violet-green Swallows. *Auk* 71: 435–442.

Conder, P. J. 1956. The territory of the Wheatear *Oenanthe oenanthe. Ibis* 98:453–459.

Cowan, J. B. 1952. Life history and productivity of a population of Western Mourning Doves in California. *Calif. Fish and Game,* 38:505–521.

Coward, T. A. 1928. *The birds of the British Isles and their eggs.* 3d ed. London: Frederick Warne and Co.

Craig, J. L. 1980. Pair and group breeding behaviour of a communal gallinule, *Porphyrio p. melanotus. Anim. Behav.* 28:593–603.

Cullen, J. M. 1954. The diurnal rhythm of birds in the Arctic summer. *Ibis* 96:31–46.

Davies, S. J. J. F. 1962. The nest-building behaviour of the Magpie-Goose *Anseranas semipalmata. Ibis* 104:147–157.

Diamond, A. W. 1975a. Breeding biology and behavior of frigate-birds *Fregata* spp. on Aldaba Atoll. *Ibis* 117:302–323.
———. 1975b. The biology of tropicbirds at Aldaba Atoll, Indian Ocean. *Auk* 92:16–30.
Dixon, K. L. 1949. The behavior of the Plain Titmouse. *Condor* 51:110–136.
Dorst, J. 1956. Etude biologique des trochilidés des hauts plateaux péruviens. *L'Oiseau et R. F. O.* 26:165–193.
———. 1957a. The Puya stands of the Peruvian high plateaux as a bird habitat. *Ibis* 99:594–599.
———. 1957b. La Vie sur les hauts plateaux andins de Pérou. *La Terre et la vie*, pp. 3–50.
Dorward, D. F. 1962. Comparative biology of the White Booby and the Brown Booby *Sula* spp. at Ascension. *Ibis* 103b:174–220.
Edson, J. M. 1943. A study of the Violet-green Swallow. *Auk* 60:396–403.
Emlen, J. T., Jr. 1938. Midwinter distribution of the American Crow in New York state. *Ecology* 19:264–275.
Emlen, S. T., and N. J. Demong. 1984. Bee-eaters of Baharini. *Natural History* 93:51–58.
Erickson, M. M. 1938. Territory, annual cycle, and numbers in a population of Wren-tits (*Chamaea fasciata*). *Univ. Calif. Publ. Zool.* 42:247–334.
Erpino, M. J. 1968. Nest-related activities of Black-billed Magpies. *Condor* 70:154–165.
Erskine, A. J. 1972. *Buffleheads.* Ottawa: Canadian Wildlife Service.
ffrench, R. P. 1967. The Dickcissel on its wintering grounds in Trinidad. *Living Bird* 6:123–140.
Fitch, F. W., Jr. 1947. The roosting tree of the Scissor-tailed Flycatcher. *Auk* 64:616.
Fleming, T. H. 1981. Winter roosting and feeding behaviour of Pied Wagtails *Motacilla alba* near Oxford, England. *Ibis* 123:463–476.
Forshaw, J. M., and W. T. Cooper. 1977. *Parrots of the world.* Neptune, N. J.: T. F. H. Publications.
Fraga, R. M. 1979. Helpers at the nest in passerines from Buenos Aires Province, Argentina. *Auk* 96:606–608.
———. 1980. The breeding of Rufous Horneros (*Furnarius rufus*). *Condor* 82:58–68.
Fredrickson, L. H. 1971. Common Gallinule breeding biology and development. *Auk* 88:914–919.
French, N. R. 1959. Life history of the Black Rosy-Finch. *Auk* 76:159–180.
French, N. R., and R. W. Hodges. 1959. Torpidity in cave-roosting hummingbirds. *Condor* 61:223.
Frith, H. J., and S. J. J. F. Davies. 1961. Ecology of the Magpie Goose *Anseranas semipalmata* Latham (Anatidae). *C.S.I.R.O. Wildlife Research* 6:91–141.
Gadgil, M. 1972. The function of communal roosts: Relevance of mixed roosts. *Ibis* 114:531–533.
Garnett, S. T. 1978. The behaviour patterns of the Dusky Moorhen, *Gal-*

linula tenebrosa Gould (Aves:Rallidae). *Australian Wildlife Research* 5:363–384.

Gaston, A. J. 1973. The ecology and behaviour of the Long-tailed Tit. *Ibis* 115:330–351.

———. 1977. Social behaviour within groups of Jungle Babblers (*Turdoides striatus*). *Anim. Behav.* 25:828–848.

———. 1978a. Social behaviour of the Yellow-eyed Babbler *Chrysomma sinensis. Ibis* 120:361–364.

———. 1978b. Ecology of the Common Babbler *Turdoides caudatus. Ibis* 120:415–432.

Gaston, T. 1987. Seabird citadels of the Arctic. *Natural History* 96(4): 54–59.

Gilliard, E. T. 1958. *Living birds of the world.* New York: Doubleday and Co.

Gochfeld, M. 1971. Notes on a nocturnal roost of Spotted Sandpipers in Trinidad, West Indies. *Auk* 88:167–168.

Goodwin, D. 1967. *Pigeons and doves of the world.* London: British Museum (Natural History).

———. 1976. *Crows of the world.* Ithaca, N.Y.: Cornell University Press.

Greig-Smith, P. W. 1979. Observations of nesting and group behaviour of Seychelles White-eyes *Zosterops modestus. Ibis* 121:344–348.

Grey of Fallodon. 1927. *The charm of birds.* New York: Frederick A. Stokes Co.

Griffin, D. R. 1974. *Listening in the dark: The acoustic orientation of bats and men.* New York: Dover Publications.

Grimes, L., and K. Darku. 1968. Some recent breeding records of *Picathartes gymnocephalus* in Ghana and notes on its distribution in West Africa. *Ibis* 110:93–99.

Groskin, H. 1945. Chimney Swifts roosting at Ardmore, Pennsylvania. *Auk* 62:361–370.

Grubb, T. C., Jr. 1973. Absence of individual distance in the Tree Swallow during adverse weather. *Auk* 90:432–433.

Gullion, G. W. 1954. The reproductive cycle of American Coots in California. *Auk* 71:366–412.

Hardy, J. W. 1963. Epigamic and reproductive behavior of the Orange-fronted Parakeet. *Condor* 65:169–199.

Harris, M. P. 1973. The biology of the Waved Albatross *Diomedea irrorata* of Hood Island, Galápagos. *Ibis* 115:483–510.

Hartley, G. I. 1917. Nesting habits of the Grey-breasted Martin. In Beebe, Hartley, and Howes 1917.

Haverschmidt, F. 1951. Notes on the life history of *Picumnus minutissimus* in Surinam. *Ibis* 93:196–200.

———. 1953. Notes on the life history of the Blood-colored Woodpecker in Surinam. *Auk* 70:21–25.

———. 1958. Notes on the breeding habits of *Panyptila cayennensis. Auk* 75:121–130.

Howard, L. 1952. *Birds as individuals.* London: Collins.

Hudson, W. H. 1920. *Birds of La Plata.* 2 vols. London: J. M. Dent and Sons.

Imber, M. J. 1976. Breeding biology of the Grey-faced Petrel *Pterodroma macroptera gouldi. Ibis* 118 : 51–64.
Immelmann, K. 1962. Beiträge zu einer vergleichenden Biologie australisher Prachtfinken (Spermestidae). *Zool. Jb. Syst.* 90 : 1–196.
Ingels, J., and J.-H. Ribot. 1983. The Blackish Nightjar *Caprimulgus nigrescens* in Surinam. *Gerfaut* 73 : 127–146.
Irving, L. 1960. Birds of Anaktuvuk Pass, Kobuk, and Old Crow: A study in Arctic adaptation. *U.S. Natl. Mus. Bull.* 217.
Jaeger, E. C. 1948. Does the Poor-will hibernate? *Condor* 50:45–46.
———. 1949. Further observations on the hibernation of the Poor-will. *Condor* 51 : 105–109.
Johnston, R. F. 1960. Behavior of the Inca Dove. *Condor* 62:7–24.
Jourdain, F. C. R. 1936. On the winter habits of the Green Woodpecker (*Picus viridis virescens*). *Proc. Zool. Soc. London,* Part 1:251–256.
Jumber, J. F. 1956. Roosting behavior of the Starling in central Pennsylvania. *Auk* 73:411–426.
Kale, H. W. II. 1962. A captive Marsh Wren helper. *Oriole,* June (unpaged reprint).
Kendeigh, S. C. 1941. Territorial and mating behavior of the House Wren. *Illinois Biol. Monogr.* 18 : 1–120.
———. 1952. Parental care and its evolution in birds. *Illinois Biol. Monogr.* 22. Urbana: University of Illinois Press.
———. 1961. Energy of birds conserved by roosting in cavities. *Wilson Bull.* 73:140–147.
Kilham, L. 1971. Roosting habits of White-breasted Nuthatches. *Condor* 73:113–114.
King, B. R. 1980. Social organization and behaviour of the Grey-crowned Babbler *Pomatostomus temporalis. Emu* 80: 59–76.
King, J. R. 1986. The daily activity period of nesting White-crowned Sparrows in continuous daylight at 65° N compared with activity period at lower latitudes. *Condor* 88 : 382–384.
Kluyver, H. N. 1950. Daily routines of the Great Tit, *Parus m. major* L. *Ardea* 38:99–135.
———. 1957. Roosting habits, sexual dominance and survival in the Great Tit. *Coldspring Harbor Symposia on Quantitative Biol.* 22:281–285.
Knorr, O. A. 1957. Communal roosting of the Pygmy Nuthatch. *Condor* 59:398.
Lack, D. 1956. *Swifts in a tower.* London: Methuen and Co.
Lancaster, D. A. 1964. Biology of the Brushland Tinamou, *Nothoprocta cinerascens. Amer. Mus. Nat. Hist. Bull.* 127:269–314.
Laskey, A. R. 1943. In Tennessee Ornithological Society 1943.
———. 1958. A winter roost of Purple Finches. *Auk* 75:475–476.
Lewis, D. M. 1982. Cooperative breeding in a population of White-browed Sparrow Weavers *Plocepasser mahali. Ibis* 124: 511–522.
Ligon, J. D. 1968. The biology of the Elf Owl, *Micrathene whitneyi. Mus. Zool. Univ. Michigan Misc. Publ.* No 136.

Ligon, J. D., and S. H. Ligon. 1979. The communal social system of the Green Woodhoopoe in Kenya. *Living Bird* 17(for 1978): 159–197.
Linsdale, J. M. 1937. *The natural history of magpies.* Pacific Coast Avifauna 25. Berkeley, Calif.: Cooper Ornithological Society.
Littlefield, C. D. 1986. Autumn Sandhill Crane habitat use in southeast Oregon. *Wilson Bull.* 98: 131–137.
Lockley, R. M. 1942. *Shearwaters.* London: J. M. Dent and Sons.
Low, J. B. 1945. Ecology and management of the Redhead, *Nyroca americana,* in Iowa. *Ecol. Monogr.* 15: 35–69.
Lunk, W. A. 1962. *The Rough-winged Swallow,* Stelgidopteryx ruficollis *(Vieillot): A study based on its breeding biology in Michigan.* Publ. Nuttall Ornith. Club 4. Cambridge, Mass.
MacDonald, S. D. 1970. The breeding behavior of the Rock Ptarmigan. *Living Bird* 9: 195–238.
Macgregor, D. E. 1950. Notes on the breeding of the Red-breasted Chat *Oenanthe heuglini. Ibis* 92: 380–383.
Maclean, G. L. 1973. The Sociable Weaver, Parts 1–5. *Ostrich* 44: 176–261.
MacRoberts, M. H., and B. R. MacRoberts. 1976. *Social organization and behavior of the Acorn Woodpecker in central coastal California.* Amer. Ornith. Union, Ornith. Monogr. 21.
Marshall, J. T., Jr. 1943. Additional information concerning the birds of El Salvador. *Condor* 45: 21–23.
Mayhew, W. W. 1958. The biology of the Cliff Swallow in California. *Condor* 60: 7–37.
Meinertzhagen, R. 1956. Roost of wintering harriers. *Ibis* 98: 535.
Meservey, W. R., and G. F. Kraus. 1976. Absence of "individual distance" in three swallow species. *Auk* 93: 177–178.
Miller, E. W. 1941. Behavior of the Bewick Wren. *Condor* 43: 81–99.
Moore, A. D. 1945. Winter night habits of birds. *Wilson Bull.* 57: 253–260.
Moreau, R. E. 1939. Numerical data on African birds' behaviour at the nest: *Hirundo s. smithii* Leach, the Wire-tailed Swallow. *Proc. Zool. Soc. London,* Ser. A, 109: 109–125.
———. 1940. Numerical data on African birds' behaviour at the nest, II: *Psalidoprocne holomelaena massaica* Neum. the Rough-wing Bank-Martin. *Ibis,* Ser. 14, 4: 234–248.
———. 1941. A contribution to the breeding biology of the Palm-Swift, *Cypselus parvus. Journ. E. Africa and Uganda Nat. Hist. Soc.* 15: 154–170.
———. 1942a. *Colletoptera affinis* at the nest. *Ostrich* 13: 137–147.
———. 1942b. The breeding biology of *Micropus caffer streubelii,* the White-rumped Swift. *Ibis,* Ser. 14, 6: 27–49.
———. 1949a. The breeding of the Paradise Flycatcher. *Ibis* 91: 256–279.
———. 1949b. The African Mountain Wagtail *Motacilla clara* at the nest. In *Ornithologie als biologische Wissenschaft,* pp. 183–191.
Moreau, R. E., and W. M. Moreau. 1937. Biological and other notes on some East African birds. *Ibis,* Ser. 14, 1: 152–174.
Morrison, D. W., and D. F. Caccamise. 1985. Ephemeral roosts and stable

patches? A radiotelemetry study of communally roosting Starlings. *Auk* 102:793–804.

Narosky, S., R. Fraga, and M. de la Peña. 1983. *Nidificación de las aves argentinas (Dendrocolaptidae y Furnariidae).* Buenos Aires: Asociación Ornitológica del Plata.

Nettleship, D. N., and T. R. Birkhead. 1985. *The Atlantic Alcidae.* London and New York: Academic Press.

Nice, M. M. 1943. Studies in the life history of the Song Sparrow, II: The behavior of the Song Sparrow and other passerines. *Trans. Linn. Soc. New York* 6:1–328.

Nice, M. M., and R. H. Thomas. 1948. A nesting of the Carolina Wren. *Wilson Bull.* 60:139–158.

Norris, R. A. 1958. Comparative biosystematics and life history of the nuthatches *Sitta pygmaea* and *Sitta pusilla. Univ. Calif. Publ. Zoöl.*: 56:119–300.

Noske, R. A. 1980. Co-operative breeding and plumage variations in the Orange-winged (Varied) Sittella. *Corella* 4:45–53.

———. 198?. The private lives of treecreepers. *Australian Nat. Hist.* 20: 419–424.

Oberlander, G. 1939. The history of a family of Black Phoebes. *Condor* 41:133–151.

Odum, E. P. 1941. Annual cycle of the Black-capped Chickadee—2. *Auk* 58:518–535.

Oniki, Y. 1970. Roosting behavior of three species of woodcreepers (Dendrocolaptidae) in Brazil. *Condor* 72:233.

Orians, G. H. 1961. The ecology of blackbird (*Agelaius*) social systems. *Ecol. Monogr.* 31:285–312.

Parmelee, D. F. 1968. Snow Bunting. *In* Bent et al. 1968.

Pearson, O. P. 1953. Use of caves by hummingbirds and other species at high altitudes in Peru. *Condor* 55:17–20.

Perry, R. 1946. *Lundy, isle of Puffins.* 2d ed. London: Lindsay Drummond.

Petersen, A. J. 1955. The breeding cycle in the Bank Swallow. *Wilson Bull.* 67:235–286.

Phillips, R. E., and H. C. Black. 1956. A winter population study of the Western Winter Wren. *Auk* 73:401–410.

Pickens, A. L. 1935. Evening drill of Chimney Swifts during the late summer. *Auk* 52:149–153.

Pitts, T. D. 1976. Fall and winter roosting habits of Carolina Chickadees. *Wilson Bull.* 88:603–610.

Preble, C. S. 1961. Unusual behavior of a House Wren. *Auk* 78:442.

Rice, D. W., and K. W. Kenyon. 1962. Breeding cycles and behavior of Laysan and Black-footed albatrosses. *Auk* 79:517–567.

Richdale, L. E. 1943. The White-faced Storm Petrel or Takahi-kare-moana, Part I. *Trans. Royal Soc. New Zealand* 73:97–115.

———. 1951. *Sexual behavior in penguins.* Lawrence: University of Kansas Press.

Ripley, D. 1942. *Trail of the money bird.* New York: Harper and Brothers.

Rowan, M. K. 1955. The breeding biology and behaviour of the Redwinged Starling *Onychognathus morio. Ibis* 97:663–705.

Rudebeck, G. 1955. Some observations at a roost of European Swallows and other birds in south-eastern Transvaal. *Ibis* 97:572–580.

Ruff, F. J. 1940. Mortality among Myrtle Warblers near Ocala, Florida. *Auk* 57:405–406.

Ruttledge, R. F. 1946. Roosting habits of the Irish Coal-Tit, with some observations on other habits. *British Birds* 39:326–333. Abstract in *Ibis* 89:380 (1947).

Schifter, H. 197?. Familie Bartvögel. In *Grzimeks Tierleben* 9:63–75. Munich: Kindler Verlag.

Scott, P., and the Wildfowl Trust. 1972. *The swans.* London: Michael Joseph.

Shaw, W. T. 1936. Winter life and nesting studies of Hepburn's Rosy Finch in Washington State. *Auk* 53:9–16, 133–149.

Sherman, A. R. 1952. *Birds of an Iowa dooryard.* Boston: Christopher Publishing House.

Short, L. 1973. Habits of some Asian woodpeckers (Aves, Picidae). *Amer. Mus. Nat. Hist. Bull.* 152:253–364.

Short, L. L., and J. F. M. Horne. 1984. Behavioural notes on the White-eared Barbet *Stactolaema leucotis* in Kenya. *Bull. British Ornith. Club* 104: 47–53.

Simons, T. K. 1985. Biology and behavior of the endangered Hawaiian Dark-rumped Petrel. *Condor* 87:229–245.

Skead, C. J. 1959. A study of the Cape Penduline Tit *Anthoscopus minutus minutus* (Shaw and Nodder). *Proc. First Pan-African Ornith. Congr.* pp. 274–288.

Skead, C. J., and G. A. Ranger. 1958. A contribution to the biology of the Cape Province white-eyes (*Zosterops*). *Ibis* 100:319–333.

Skutch, A. F. 1935. Helpers at the nest. *Auk* 52:257–273.

———. 1940. Social and sleeping habits of Central American wrens. *Auk* 57:293–312.

———. 1944. Life-history of the Prong-billed Barbet. *Auk* 61:61–88.

———. 1945a. The migration of Swainson's and Broad-winged hawks through Costa Rica. *Northwest Sci.* 19:80–89.

———. 1945b. Life history of the Blue-throated Green Motmot. *Auk* 62: 489–517.

———. 1948a. Life history of the Golden-naped Woodpecker. *Auk* 65: 225–260.

———. 1948b. Life history of the Olivaceous Piculet and related forms. *Ibis* 90:433–449.

———. 1953. Life history of the Southern House Wren. *Condor* 55:121–149.

———. 1954. *Life histories of Central American birds* [I]. Pacific Coast Avifauna 31. Berkeley, Calif.: Cooper Ornithological Society.

———. 1958. Roosting and nesting of araçari toucans. *Condor* 60:201–219.

———. 1960. *Life histories of Central American birds II.* Pacific Coast Avifauna 34. Berkeley, Calif.: Cooper Ornithological Society.

———. 1961. The nest as a dormitory. *Ibis* 103a:50–70.

———. 1967. *Life histories of Central American highland birds.* Publ. Nuttall Ornith. Club 7. Cambridge, Mass.
———. 1968. The nesting of some Venezuelan birds. *Condor* 70:66–82.
———. 1969a. A study of the Rufous-fronted Thornbird and associated birds. *Wilson Bull.* 81:5–43, 123–139.
———. 1969b. *Life histories of Central American birds III.* Pacific Coast Avifauna 35. Berkeley, Calif.: Cooper Ornithological Society.
———. 1972. *Studies of tropical American birds.* Publ. Nuttall Ornith. Club 10. Cambridge, Mass.
———. 1973. *The life of the hummingbird.* New York: Crown Publishers.
———. 1976. *Parent birds and their young.* Austin: University of Texas Press.
———. 1980. *A naturalist on a tropical farm.* Berkeley: University of California Press.
———. 1981. *New studies of tropical American birds.* Publ. Nuttall Ornith. Club 19. Cambridge, Mass.
———. 1983a. *Birds of tropical America.* Austin: University of Texas Press.
———. 1983b. *Nature through tropical windows.* Berkeley: University of California Press.
———. 1987. *Helpers at birds' nests: A worldwide survey of cooperative breeding and related behavior.* Iowa City: University of Iowa Press.
Skutch, A. F., and D. Gardner. 1985. *Life of the woodpecker.* Santa Monica, Calif.: Ibis Publishing Co.
Smith, S. M. 1972. Roosting aggregations of Bushtits in response to cold temperatures. *Condor* 74:478–479.
Snow, D. W. 1961. The natural history of the Oilbird, *Steatornis caripensis* in Trinidad, W. I., Part I: General behavior and breeding habits. *Zoologica* (New York Zool. Soc.) 46:27–48.
———. 1962. Notes on the biology of some Trinidad swifts. *Zoologica* (New York Zool. Soc.) 47:129–139.
Staebler, A. E. 1941. The number of feathers in the English Sparrow. *Wilson Bull.* 53:126–127.
Stickel, D. W. 1964. Roosting habits of Red-bellied Woodpeckers. *Wilson Bull.* 76:382–383.
Still, E., P. Monaghan, and E. Bignall. 1987. Social structuring at a communal roost of Choughs *Pyrrhocorax pyrrhocorax. Ibis* 129:398–403.
Stone, R. H. 1950. Roosting Brown Creepers, *Certhia familiaris. Auk* 67:391.
Stonehouse, B. 1953. The Emperor Penguin *Aptenodytes fosteri* Gray, I: Breeding behaviour and development. Falkland Islands Dependencies Survey. *Sci. Reports,* No. 6:1–33. London: H. M. Stationery Office.
———. 1960. The King Penguin *Aptenodytes patagonica* of South Georgia, I: Breeding behaviour and development. Falkland Islands Dependencies Survey. *Sci. Reports,* No. 23:1–81. London: H. M. Stationery Office.
———. 1962. The tropicbirds (genus *Phaethon*) of Ascension Island. *Ibis* 103b:124–161.

Summers-Smith, D. 1958. Nest-site selection, pair formation and territory in the House-Sparrow *Passer domesticus*. *Ibis* 100:190–203.

Swift, J. J. 1959. Le Guêpier d'Europe *Merops apiaster* L. en Camargue. *Alauda* 27:97–143.

Tennessee Ornithological Society. 1943. How birds spend their winter nights: A symposium. *Migrant* 14:1–5.

Terrill, L. McI. 1943. Nesting habits of the Yellow Rail in Gaspé County, Quebec. *Auk* 60:171–180.

Thomas, B. T. 1986. The behavior and breeding of adult Maguari Storks. *Condor* 88:26–34.

Thomas, R. H. 1946. A study of Eastern Bluebirds in Arkansas. *Wilson Bull.* 58:143–183.

Tomback, D. F. 1978. Pre-roosting flight of the Clark's Nutcracker. *Auk* 95:554–562.

Trost, C. H. 1972. Adaptations of Horned Larks (*Eremophila alpestris*) to hot environments. *Auk* 89:506–527.

Tuck, L. M. 1960. *The murres: Their distribution, populations, and biology.* Ottawa: Canadian Wildlife Service.

Van Someren, V. G. L. 1956. *Days with birds: Studies of habits of some East African species.* Fieldiana: Zoology 38. Chicago: Chicago Natural History Museum.

Verner, J. 1961. Nesting activities of the Red-footed Booby in British Honduras. *Auk* 78:573–594.

———. 1965. Breeding biology of the Long-billed Marsh Wren. *Condor* 67:6–30.

Walkinshaw, L. H. 1953. Notes on the Greater Sandhill Crane (*Grus canadensis tabida*). *Auk* 70:204–205.

Ward, P., and A. Zahavi. 1973. The importance of certain assemblages of birds as "information-centers" for food finding. *Ibis* 115:517–534.

Warham, J. 1957. Notes on the roosting habits of some Australian birds. *Emu* 57:78–81.

———. 1961. The birds of Raine Island, Pandora Cay and Murray Island Sandbank, North Queensland. *Emu* 61:77–93.

———. 1974a. The breeding biology and behaviour of the Snares Crested Penguin. *Journ. Royal Soc. New Zealand* 4:63–108.

———. 1974b. The Fiordland Crested Penguin *Eudyptes pachyrhynchus*. *Ibis* 116:1–27.

Weatherhead, P. J., and D. J. Hoysak. 1984. Dominance structuring of a Red-winged Blackbird roost. *Auk* 101:551–555.

Weatherhead, P. J., S. G. Sealy, and R. M. R. Barclay. 1985. Risks of clustering in thermally-stressed swallows. *Condor* 87:443–444.

Weller, M. W., I. C. Adams, Jr., and B. J. Rose. 1955. Winter roosts of Marsh Hawks and Short-eared Owls in central Missouri. *Wilson Bull.* 67:189–193.

Welter, W. A. 1935. The natural history of the Long-billed Marsh Wren. *Wilson Bull.* 47:3–34.

Wetmore, A. 1936. The number of contour feathers in passeriform and related birds. *Auk* 53:159–169.

———. 1945. Sleeping habits of the Willow Ptarmigan. *Auk* 62:638.

White, F. N., G. A. Bartholomew, and T. R. Howell. 1975. The thermal significance of the nest of the Sociable Weaver *Philetairus socius:* Winter observations. *Ibis* 117:171–179.

Williams, J. G. 1951. *Nectarinia johnstoni:* A revision of the species, together with data on plumage, moults and habits. *Ibis* 93:579–595.

Williams, L. 1941. Roosting habits of the Chestnut-backed Chickadee and the Bewick Wren. *Condor* 43:274–285.

———. 1947. A Winter Wren roost. *Condor* 49:124.

Wynne-Edwards, V. C. 1962. *Animal dispersion in relation to social behaviour.* Edinburgh and London: Oliver and Boyd.

Yom-Tov, Y. 1979. The disadvantage of low position in colonial roosts: An experiment to test the effect of droppings on plumage quality. *Ibis* 121:331–333.

Yom-Tov, Y., A. Imber, and J. Otterman. 1977. The microclimate of winter roosts of the Starling *Sturnus vulgaris. Ibis* 119:366–368.

Young, A. M. 1971. Roosting of a Spotted Antbird (Formicariidae: *Hylophylax naevioides*) in Costa Rica. *Condor* 73:367–368.

Zahavi, A. 1971a. The function of pre-roost gatherings and communal roosts. *Ibis* 113:106–109.

———. 1971b. The social behaviour of the White Wagtail *Motacilla alba alba* wintering in Israel. *Ibis* 113:203–211.

Zammuto, R. M., and E. C. Franks. 1981. Environmental effects on roosting behavior of Chimney Swifts. *Wilson Bull.* 93:77–84.

Zonov, G. B. 1967. [On winter roosting of Paridae in Cisbaikal.] *Ornitologiya* 8:351–354. (Russian.) Abstract in *Ibis* 110:391 (1968).

Index